BASIC molecular spectroscopy

P.A. Gorry, BA, PhD

Lecturer in Physical Chemistry,
Department of Chemistry,
University of Manchester

Butterworths
London . Boston . Durban . Singapore . Sydney . Toronto . Wellington

First published 1985

British Library Cataloguing in Publication Data

Gorry, P.A.
BASIC molecular spectroscopy.——(Butterworths BASIC series)
1. Molecular spectroscopy——Data processing
2. Basic (Computer program language)
I. Title
535.8′4′028542 QC454.M6

ISBN 0–408–01553–5

Library of Congress Cataloging in Publication Data

Gorry, P. A.
BASIC molecular spectroscopy.

(Butterworths BASIC series)
Bibliography: p.
Includes index.
1. Molecular spectroscopy—Computer programs.
2. Basic (Computer program language) I. Title.
II. Series.
QD96.M65G67 1985 539′.6′028 85–14992
ISBN 0–408–01553–5

Typeset by Mid-County Press, 2a Merivale Road, London SW15 2NW
Printed and bound in England by Anchor-Brendon Ltd., Tiptree, Essex

B roscopy

Butterworths BASIC Series includes the following titles:

BASIC aerodynamics
BASIC economics
BASIC hydraulics
BASIC hydrology
BASIC interactive graphics
BASIC investment appraisal
BASIC materials studies
BASIC matrix methods
BASIC mechanical vibrations
BASIC numerical mathematics
BASIC soil mechanics
BASIC statistics
BASIC stress analysis
BASIC theory of structures
BASIC thermodynamics and heat transfer

To Charmian

Preface

The last decade has seen an explosive increase in the use of computers in all aspects of life. It is now common to find computers in the home, the school, in work and in the laboratory. For many years computers were the prerogative of large institutions but all that has been changed by the emergence of the ubiquitous 'micro'. The ready availability of cheap computing power that, only a few years ago, would have cost many thousands of pounds, has created the need for a new type of textbook—one that utilizes this new-found computer power to extend and enlarge upon the more traditional methods of teaching scientific material.

This book has the aim of providing an introduction to molecular spectroscopy whilst making extensive use of computer programs written in BASIC to provide practical examples and illustrations of the topics under consideration. The book is primarily aimed at first- and second-year undergraduates in Chemistry and Physics taking an introductory course in molecular spectroscopy, however, it should also serve as a source book for more advanced undergraduates and those who use molecular spectroscopy as a practical tool.

The book contains a large number of well-documented BASIC programs which form the basis for a library of practical problem-solving routines. BASIC was chosen as the language for this book, and indeed for this whole series, since it is simple to write, easy to use and almost universally available on microcomputers. A further important consideration is the ease with which programs can be written, tested, modified and run in an *interactive* mode. This results in much faster program development—albeit at the cost of less elegance—than is possible with computer languages such as FORTRAN or PASCAL.

It is expected that readers will want to modify the programs presented here to suit their own particular needs and, indeed, each chapter is accompanied by a number of problems designed to explore spectroscopic and computational aspects of the material in more depth. Although the majority of problems are closely related to the programs and text itself, some are given that provide the student with more challenging material and the chance to develop certain aspects of the subject to a higher level. To this end the reader is also referred to other books in the series, such as *BASIC Numerical Mathematics* and *BASIC Matrix Methods*, for more advanced numerical techniques.

In a subject as large as molecular spectroscopy, a book of this length presents the author with a difficult dilemma. One can either attempt to treat all areas equally, albeit at a superficial level, or one is forced to restrict the scope so as to achieve a reasonable coverage of selected topics. The second approach has been adopted here for two reasons; first, it provides for greater coherence and a more satisfying level of exposition. Secondly, the tremendous strides in instrumentation and the advent of lasers in particular, has meant that high resolution spectroscopy is now the norm rather than the exception. It thus becomes imperative that any discussion of modern spectroscopy should concern itself with the finer molecular interactions that are responsible for the complex spectra revealed at high resolution.

The contents of this book are concerned solely with spectra arising from the interaction of the electric field component of light with molecular or electronic motions. The chapters deal successively with programming, fundamentals of molecular quantum mechanics and light absorption, rotational spectroscopy, vibrational spectroscopy, and finally Raman and electronic spectroscopy. The lack of space has precluded a discussion of molecular symmetry or magnetic phenomena and, although this is to be regretted, it is hoped that the greater detail afforded to the other topics will more than compensate.

PAG
1985

Contents

Chapter 1

Introduction to BASIC

1.1 What is BASIC?

The name BASIC is an acronym for Beginner's All-purpose Symbolic Instruction Code, and was developed at Dartmouth College, USA, as a general purpose computer language aimed at newcomers to computing. It is designed to be very easy to use and represents one of the simpler introductions to programming languages. In particular it is well suited to a 'conversational' or 'interactive' style of computing in which the user supplies information to the computer in response to questions issued by the program.

A BASIC program is a list of instructions to be obeyed by the computer. Each instruction is rather simple and usually contains words that are readily comprehensible even to a beginner—such as PRINT, INPUT, IF, THEN, END.... These, however, are far too difficult for a computer to act on directly and BASIC usually operates via an 'interpreter'. This is a program whose task is to take the short BASIC instructions and to convert them into myriad machine level instructions which are then acted upon immediately. The interpreter is normally 'invisible' to the user, although it's always there in the background ceaselessly carrying out its task.

1.2 Why BASIC?

The chief attraction of BASIC—its simplicity—is also its main weakness. There are several other computer languages such as FORTRAN, ALGOL and PASCAL which are much more powerful and offer a wider choice of instructions to the programmer. The price to be paid for the extra flexibility is that the time spent learning the language is much greater. Further, these languages are not, in general, orientated towards a conversational style of computing. As a result the time taken to run, test, modify and correct a program is also much longer. A further point to note is, that normally, the better the language the more space it takes up in the computer—and the less room it leaves for your programs. These factors weigh against the use of one of the more powerful languages and in favour of BASIC. This, coupled with the fact that BASIC is standard on nearly all microcomputers, makes it the best all-round choice.

1.3 Programming in BASIC

1.3.1 *Programming practice*

This book is not intended as an instruction manual for the BASIC language itself, although the majority of this chapter is devoted to important elements of the language. Those Readers who require such a work are referred to References 1–3, or one of the many similar books that are now available on this widely used language. An important task for the book is, however, to help the student learn programming techniques in general and BASIC in particular—by using it to solve real problems in molecular spectroscopy.

Programming is a skill which has to be learnt and like most skills, there are a few guidelines which will go a long way towards transforming the beginner into an expert practitioner. A further requirement is practise—lots of it.

The test of a good program is whether or not someone who didn't write it can easily understand how it functions (I take for granted that it already passes the test of giving the right answers!). Some programs resemble the stereotype of the absent minded professor—they wander all over the place, use rambling phrases with little or no explanation and look totally disorganized—following their logic is a feat of mental gymnastics. It is hoped that by working through the programs and exercises in this book the reader will avoid most of the pitfalls and develop a sound programming style.

It is a fact that BASIC tends to encourage sloppy and disorganized programming. There are several reasons for this; first, because it is so easy to write, people tend to 'leap straight in' and start programming before fully specifying what the program is to do. The code (program instructions) then grows in an organic way, sprouting branches and appendages in a completely disorganized manner. A second reason arises from a limitation in BASIC itself—its lack of *independent* subroutines with their own internal variables. In consequence there is less reason to delegate tasks to subroutines, resulting in a less structured program and poorer clarity. Thirdly, standard BASIC only allows two letters for variable names; TEMPERATURE and PRESSURE are much clearer than TE and PR. Finally, BASIC has rather poor *conditional instructions* which results in disjointed program flow with GOTO statements used to leap over blocks of code.

An example of the difference in approach between BASIC and PASCAL is that 'GOTO' is widely used in the former and is considered an admission of failure in the latter (I have used one PASCAL compiler that treated GOTOs as an error!).

There are some versions of BASIC, usually called Structured or Extended BASIC, which overcome many or all of these limitations. However, in order to make the programs in this book as widely applicable as possible, only standard BASIC will be used here.

In an effort to promote sound programming practice we shall adopt a structured approach to most of the problems and programs presented in this book. It is a somewhat artificial constraint for short programs and often results in a slightly slower execution speed, however the gain in clarity usually more than compensates.

1.3.2 *Structured programming*

The essence of structured programming is to break down the task at hand into a series of sub-tasks. The main program then calls these sub-tasks in the correct order. Each sub-task may, in itself, then be divided into smaller units (sub-sub-tasks). In this way a complex problem is broken down into a hierarchy of successively simpler tasks, until each one is small enough to produce easy coding of the instructions. This approach also enables each of the small routines to be tested separately, greatly facilitating the 'debugging' stage of program development.

The structured approach is evident in everyday life, for instance, the instructions

Go to the shops
Buy some bread
Return home

represent the top layer of a structured program. The first instruction summons up a whole host of other problems; where are the shops?, how shall I get there?, what should I take?... Having answered these questions we then need a whole series of other 'routines'; walk to the car, get in.... Even walking requires a complex set of instructions to be sent to the legs!

By adopting a structured approach it is simple to see what overall task has been set. It would be a very different matter if we had been presented with one huge list of instructions telling us what to do with our arms and legs. Indeed we would consider it madness to be given instructions in this format—it's a pity that so many programmers don't seem to realize this.

1.4 The elements of BASIC

BASIC, like most widely used languages, is not uniquely defined and static. There are many different variants so no two versions are guaranteed to run the same program. However, the many versions all share a common core and it is this core that we shall be using throughout the book. In particular we shall not make use of long variable names, true 'independent' subroutines, extended conditional statements (such as IF...THEN...ELSE and CASE), or matrix assignments. Some of the programs in this book would benefit considerably from using enhanced versions of BASIC and readers are encouraged to modify the programs to make full use of such features.

The elements of BASIC fall into six main areas: definitions, input, output, calculation, program control and subroutines. Before looking at each of these in turn we should first say a few words about variable types.

BASIC supports two main types of variable; numbers and strings. In general, these variables can have two characters in their name—the first *must* be a letter whilst the second can be a number or letter. Thus, valid variable names are X, PR, Z2 but not 4Y or MOL—although many versions allow more than two characters and only use the first two. Strings are lists of characters and string variables are designated by a $ sign at the end of the name, e.g. A$,P1$.

Variables are given values by assignment statements using the = symbol.

```
X2 = 3.5          (numbers)
A$ = "FRED"       (strings)
```

Although some versions also distinguish INTEGER (4, 28...), as opposed to REAL (1.2, 3.7...), numbers this will not be used here.

1.4.1 *Definition statements*

A BASIC program is a list of instructions to be obeyed by the computer. The overall sequence of a program is defined by 'line numbers' attached to each statement. BASIC automatically orders the list into ascending line numbers so that typing in

```
100 PRINT"HELLO THERE"
300 PRINT"TODAY?"
200 PRINT"HOW ARE YOU"
```

Would produce the program

```
100 PRINT"HELLO THERE"
200 PRINT"HOW ARE YOU"
300 PRINT"TODAY?"
```

As well as defining the order in which the program statements will be obeyed, line numbers also provide labels with which to direct program flow, using GOTO and GOSUB commands. A useful hint on line numbering is to leave plenty of room between successive line number (gaps of 10 or 100) so that there is plenty of space to insert statements at a later date.

A very important 'definition' statement, one which should feature prominently in all good programs, is the REM statement. This permits comments to be inserted into a program. The computer makes no attempt to interpret a REM statement—it passes directly to the next line. REM statements make it easy to understand what a program is doing and greatly facilitate updating and improving the program at a later date. It is surprising just how abstruse 'un-commented' computer code can look just a few weeks after writing it. In most versions of BASIC, REM statements can be added to the end of a line as well as occupying a line of their own.

```
100 REM      On its own line
110 X=1 : REM       At the end of a line.
```

Both types of REM statement will be used here.

In mathematics it is frequently desirable to use subscripted variables. We may wish to refer to particular bond lengths in a molecule, containing N bonds, by R_i ($i=1$ to N). In BASIC we use an *array* to hold this information. An *element* of the array would be referred to as R(1), R(5), or R(I)... where I has a value between 1 and N. We use array elements as normal variables.

```
100 R(3) = 1.7
110 D = R(1)+R(2)
```

It is possible to have arrays which are two dimensional. If we wanted to store the X and Y coordinates of each atom we could use X(I) and Y(I), but equally we could use a single array C(I,J) where I=1 to N and J=1 to 2. The I index picks the atom and the J index picks the X or Y coordinate. Most BASICs allow more than just two dimensions—but it is comparatively rare to need this. However, all computers need to know how much space they must set aside for the array, so it is necessary to 'dimension' the array before using it. For this reason arrays are usually dimensioned at the beginning of the program.

```
100 REM Program start
110 N=20
120 DIM R(100),C(100,2)
130 DIM X(N)
```

Line 130 is an example of *dynamic* array allocation, in which the array size is set by a variable in the program. Most versions of interpreted BASIC allow this, although compiled versions may not. However, arrays cannot be re-dimensioned (even to the same size) so it is important to ensure that the DIM statement is only obeyed once. A final point to note is that arrays are numbered from element 0 in BASIC, thus R(100) in a DIM statement actually creates an array with 101 elements in it. This isn't important unless you use an array like Z(2,8000) to store two pieces of information for each of 8000 objects. In fact you will have set aside 3×8001 elements rather than the 2×8000 needed—a fact that may be important if memory space is limited.

A final definition statement which you may require is the DEF FN statement. This allows one to define simple functions beyond the ones supplied in BASIC as standard (SIN, COS, EXP etc.). The rules governing the DEF FN statement tend to vary between different versions and some offer greater flexibility than others. The standard adopted here is that functions can only be of a single variable, they must occupy a single line of code and their names must conform to the two character limit for variables.

```
 90 REM Hyperbolic sine
100 DEF FN HS(X) = 0.5 * (EXP(X)-EXP(-X))
100 Z=FN HS(0.25)
```

Function definitions can be placed anywhere in the program but it is common to group them together either at the beginning or the end of the program. Also some versions permit function re-definition but this is generally poor programming practice.

1.4.2 *Input*

Most programs require information which varies from run to run. Since BASIC is a conversational language this information is often supplied by the user when the program is running, via the INPUT command. BASIC also provides the READ command to access data stored internally in DATA statements. If large quantities of information are required most systems will store this on magnetic disc or tape and read it from there. Unfortunately there is no single, standard disc operating system so no reference will be made to such 'external' storage in this book. In fact most of the programs presented here require only minimal input so this is not a serious limitation.

It is fundamental to good interactive programming that data input should be as informative and foolproof as possible. By the first we mean that the program should clearly indicate what information is being asked for at each input stage. The second is much more difficult and requires extensive internal checks to see if the data are valid. For instance, if a bond length (in angstroms) is asked for then a number less than 0.5 or greater than 4 would generally be invalid and we could refuse to accept such a number.

```
100 PRINT"BOND LENGTH = ";
200 INPUT BL
300 IF BL>0.5 AND BL<4.0 THEN GOTO 600
400 PRINT"INVALID BOND LENGTH"
500 GOTO 100
600 REM Rest of program
```

This little program illustrates the way in which the user is prompted for input (line 100) and then the value is checked to see if it is sensible (line 300). An invalid number results in the question being asked again. The program could be improved by reminding you of the units (angstroms) being used and by explicitly telling you of the allowed range if the first try is invalid. The programs in this book will always prompt the user about the information required but, in general, no attempt will be made to render them 'foolproof'. This is a direct result of trying to keep the program lengths to a minimum. If the programs are to be used by inexperienced people then it would be desirable to include greater input validation.

Most versions of BASIC allow for lines 100 and 200 to be combined in the one INPUT statement and that will be assumed throughout the book.

```
100 INPUT"BOND LENGTH = ";BL
```

The READ statement operates rather like the INPUT statement but it looks internally in the program for the information rather than asking the user. For this reason no prompting is required. The information is stored in the program in DATA statements which are read by the program in the order they occur—so it is essential to ensure that the DATA and READ statements match each other exactly.

```
100 READ M,R,F : REM mass, bond length, frequency
 :
 :
1000 DATA 35, 1.4,17.2
```

1.4.3 *Output*

The most important part of a program, as far as the user is concerned, is the output. If the output is poor and confusing then the program is really of little use—regardless of how correct its calculations are. In BASIC, output is controlled by the PRINT statement, which has the general format:

```
line number  PRINT list
```

The PRINT command is used to output fixed values, variables and strings. The 'list' can be a mixture of all of these and is very flexible.

```
100 PRINT "TEXT EXAMPLE"
110 PRINT X
120 PRINT "THREE VARIABLES",A,B,C
```

Line 120 uses commas to produce tabulation of approximately 14 spaces wide. The exact width varies between the different versions of BASIC. Each of the PRINT statements will issue a 'line-feed' at the end of its output, ensuring that the next PRINT will start on a new line. The tabulation and line-feed can be cancelled by using a semi-colon instead of a comma.

```
100 PRINT "PART ONE ";
110 PRINT "PART TWO ON THE SAME LINE"
120 PRINT A;B;C
```

The semi-colon at the end of line 100 forces the next PRINT to continue on the same line. In line 120 the values in A, B and C will be printed with *no* spaces between them. This would be very difficult to read and a much better version of line 120 would be:

```
120 PRINT "A = ";A;"B = ";B;"C = ";C
```

In this last case each of the variables would be preceded by an identifier and they would be separated by spaces. Line 120 is now a much clearer output statement and a considerable improvement on the first version. If line 120 occurred in a loop, however, it would produce straggling columns, since the position of B would depend on the number of digits output for A. Most BASICs don't provide formatting routines that allow one to specify the number of digits to be output so you must write your own. An alternative is to use the TAB( ) function to ensure that all the output starts at specified places.

```
100 PRINT X;TAB(10);Y;TAB(20);Z
```

Many computers also provide extensive screen format and positioning commands which can be used to produce very versatile output—unfortunately there are no standard commands for this so they will not be used in this book.

It is good programming practice to output all the important information the program is using—this greatly increases the chance of inadvertent data input errors being noticed. The PRINT statement is also the best 'debugging' tool available in BASIC. A liberal sprinkling of them during the testing stages will save a lot of time later on. The superfluous ones can then be removed when you are sure the program is operating correctly.

1.4.4 *Calculation*

The heart of a program is made up of a series of statements that perform mathematical manipulations. These statements usually look very similar to their corresponding mathematical expression. A major difference is that the '=' symbol in BASIC really stands for 'give the variable on the left of it the value of the expression on the right of it'.

```
100 Z=2
110 X=Y*Y
120 N=N+1
```

Line 120 shows a commonly used assignment which increments the value of N by one. This clearly does not correspond to the *mathematical* statement N=N+1.

The arithmetic operators of: raising to a power, multiplying and dividing, adding and subtracting are all provided in BASIC. They follow a hierarchy which determines the order in which they are obeyed when evaluating a BASIC statement. The order, in decreasing importance, is

(i) ^ raise to a power
(ii) *, / multiply and divide (equal importance)
(iii) +, − add and subtract (equal importance)

It is important to remember this order otherwise unexpected results can occur. For instance the BASIC line

```
100 X = A/B+C
```

will result in the value (A/B)+C—rather than A/(B+C). It is better to make generous use of parentheses than to risk an incorrect statement.

Expression	*Incorrect*	*Correct*
$r.\cos(a)$	RCOS(A)	R*COS(A)
$\frac{1+e^x}{1-e^x}$	1+EXP(X)/1−EXP(X)	(1+EXP(X))/(1−EXP(X))
$y^{a/b}$	Y^A/B	Y^(A/B)

No distinction between integers and real numbers (those with a decimal point) will be made in this book, although some versions provide integer variables. A point to note is that the trigonometric functions SIN, COS, TAN all expect their arguments to be in *radians*, not degrees.

1.4.5 *Program control*

Under normal conditions the statements in a program are carried out in the sequence defined by the line numbers. However, it is comparatively rare that one can write a complete program in such a linear manner. One normally needs to incorporate several alternative sections of code into a program—what section is obeyed depends on information supplied by the user at run time. It is also common to need to repeat a section of code many times, or indeed the whole program. BASIC provides simple statements to achieve these aims.

The simplest statement is the unconditional: GOTO *line number*. This simply causes the program to jump to the stated line number and continue processing. A more important statement is the conditional IF... THEN. It has the general form:

line number IF expression1 conditional operator expression2 THEN line number/statement

The conditional operators supported are:

```
=    equal to
<>   not equal to
<    less than
>    greater than
<=   less than or equal to
>=   greater than or equal to
```

Simple examples would be

```
100 IF X<10 THEN 200
110 IF X>20 THEN GOTO 200
120 IF X=15 THEN Z=12
```

In line 110 we follow the THEN by the *statement* GOTO 200. This tends to make the code clearer and will be adopted throughout. The statements in 100 and 110 redirect the program flow if X has a value outside the 10–20 range. In fact these two lines can be amalgamated by using *logical* operators such as AND, OR, NOT.

```
100 IF X<10 OR X>20 THEN GOTO 200
```

If a variable (or expression) can yield integer parts of 1, 2, 3 ... and different action is required for each value there is a more efficient way of redirecting the code than using multiple IF statements. This is the ON … GOTO statement, with the general form:

line number ON variable/expression GOTO line number1, line number2, …

This is especially useful for dealing with 'menu' selection, where many choices are on offer.

A very important facility in program control is the 'loop' statement. This allows a section of code to be repeated a specified number of times. This command is particularly useful when used in conjunction with subscripted variables. It has the general form:

line number FOR variable=expression1 TO expression2 [STEP expression3]

and the loop is ended with

line number NEXT variable

The step part is optional and takes the value of 1 if it is omitted. As an example suppose we wanted to fill the first 10 elements of an array X with the values 1^2, 2^2, 3^2, … we could write:

```
100 FOR I=1 to 10
110 X(I)=I^2
120 NEXT I
```

One should avoid leaping into or out of loops using GOTO statements. In some cases this can be immediately fatal to the program, in others it may be storing up trouble or confusion for later on. It is also poor programming to put a complicated expression in the loop control statement since it will be re-evaluated each loop cycle.

1.4.6 *Subroutines*

Ideally subroutines are just what their name suggests 'sub-routines'—that is simple tasks that are needed by the main program. In more sophisticated languages like PASCAL or FORTRAN these sub-routines are independent 'mini-programs' with their own variables, loops, tests, etc. They communicate to the main program via a list of variables defined at the head of the subroutine. This independence means that the subroutines can be debugged separately from the main program and popular ones can be incorporated into several different programs.

Unfortunately BASIC offers none of this independence and it is more accurate to think of a subroutine as a special GOTO statement—one that remembers where it came from. Subroutines are called using the GOSUB statement which has the general form:

line number GOSUB line number

All variables in BASIC are 'global'—that is, if X1 occurs in a subroutine it is the *same* variable as X1 in the main program. This makes transportable subroutines rather difficult since one must always ensure that the subroutine and main program don't inadvertently use the same variable name for different purposes.

The end of the subroutine is signalled by a RETURN statement. This instructs the program to return to the statement *following* the GOSUB command. Subroutines can call other subroutines but they may not call themselves—or another subroutine which does so. Subroutines that call themselves are called *recursive* and are allowed in higher languages such as PASCAL. Some extended BASICs permit recursion but it is far from standard.

ON...GOSUB operates in the same way as the ON...GOTO described in Section 1.4.5 and, like loops, one should never jump in to or out of subroutines using GOTO statements.

Subroutines will feature prominently throughout this book although, because of the limitations of BASIC, they are sometimes artificially introduced. They do however lead to more readable programs. For instance, a main program can look like:

```
100 REM GET INPUT
110 GOSUB 1000
120 REM PERFORM CALCULATIONS
130 GOSUB 2000
140 REM OUTPUT RESULTS
150 GOSUB 3000
160 END
```

where each of the subroutines starting at 1000, 2000, 3000 perform the stated tasks.

1.5 Different computers

Most microcomputers offer an extended version of BASIC compared with the one used in this book. The most common extensions are long variable names, true subroutines, more powerful conditional statements, matrix routines and graphical commands. Unfortunately, there are as many extensions as there are different micros so we must be content with a standard version here.

One area which greatly improves the comprehensibility of programs is the use of graphical display. Some of the programs here use the PRINT command to produce simple 'graphical' output and these would benefit greatly by being converted to true high resolution graphics.

The reader is encouraged to augment the programs presented

throughout this book with any special features available in their version of BASIC—although if you intend using the programs on more than one type of computer you will have to look carefully at their compatibility.

1.6 References

1. Alcock, D., *Illustrating BASIC*, Cambridge University Press (1977).
2. Gottfried, B.S., *Programming with BASIC—Schaum's Outline Series*, McGraw-Hill (1975).
3. Monro, D.M., *Interactive Computing with BASIC*, Arnold (1974).

Chapter 2

The quantum treatment of molecules

ESSENTIAL THEORY

Understanding the spectra of atoms and molecules has been one of the great challenges, and success stories, in physical science this century. The fact that many atomic and molecular spectra consist of discrete lines is not expected on the basis of classical physics and it has been a triumph of quantum mechanics to explain this observation.

Quantum mechanics places severe limitations on the allowed energy states of atomic and molecular systems and determines the allowed transitions between them. The rest of this book will examine how the principles of quantum mechanics can be used to explain molecular spectra and how this in turn leads to important structural information on the molecules involved.

2.1 The Schrödinger equation

An essential prerequisite to understanding electronic, vibrational or rotational molecular spectra is to have an expression for the energy levels of the system under study. The observed spectrum is produced when the system changes between these allowed states. The energy levels are obtained by solving the *time-independent* Schrödinger wave equation for the molecular motions under consideration.

It is best to consider the Schrödinger equation as the fundamental starting point in quantum mechanics—just as Newton's laws are in classical mechanics. We use the Schrödinger equation because it works!

This chapter will present the important aspects of quantum mechanics necessary for an understanding of molecular spectroscopy. Although it is self-contained it is not intended that this chapter should provide a comprehensive account of molecular quantum theory—even at the introductory level—and little in the way of formal solutions will be provided. Those readers not familiar with the material in this chapter should consult one of References 1–3 for more detail.

The time-independent Schrödinger equation has the form

$$\hat{H}\psi = E\psi \tag{2.1}$$

This deceptively simple equation holds the key to calculating the allowed energy states of molecular systems. We shall now look at each of its parts in turn.

2.1.1 *The hamiltonian operator, $\hat{H}$*

The starting point for solving any energy level problem is to construct the hamiltonian operator for the system. This is then used to determine the *quantized* energy levels. The hamiltonian operator is easily constructed provided we know the *classical* energy of the system. The first step is to write down the total energy—treating it in a classical way.

$$H = T + V \tag{2.2}$$

where H is the total energy, T is the kinetic energy and V the potential energy.

The operator $\hat{H}$ is then constructed using a simple set of rules; we change *variables* to *operators* in the manner given by *Table 2.1*.

Table 2.1 Rules for converting classical variables into quantum mechanical operators

	Variable	*Operator*
Coordinate	x, y, z	x, y, z
Linear momentum	P_x, P_y, P_z	$-i\hbar\frac{\partial}{\partial x}, -i\hbar\frac{\partial}{\partial y}, -i\hbar\frac{\partial}{\partial z}$
Angular momentum	L_x	$-i\hbar\left(y\frac{\partial}{\partial z}-z\frac{\partial}{\partial y}\right)$
	L_y	$-i\hbar\left(z\frac{\partial}{\partial x}-x\frac{\partial}{\partial z}\right)$
	L_z	$-i\hbar\left(x\frac{\partial}{\partial y}-y\frac{\partial}{\partial x}\right)$

where $\hbar = h/2\pi$ and $i = (-1)^{\frac{1}{2}}$

It should be noted here that the ^ symbol above a letter is used to denote an *operator* as opposed to a classical variable. This convention will be used throughout the book.

The rules may seem esoteric but it is beyond the scope of this book to describe how they arise. However, by following the prescription above we can easily construct the hamiltonian operator for various systems of interest.

For instance consider the case of a particle moving in one dimension only, with no external forces acting upon it. Classically we write:

$$T = \frac{1}{2} m\dot{x}^2 = \frac{1}{2m}(m\dot{x})^2 = \frac{1}{2m} P_x^2$$

$$V = 0, \qquad \text{where } \dot{x} = \frac{\mathrm{d}x}{\mathrm{d}t}$$

We must now replace P_x by $-i\hbar(\partial/\partial x)$

$$P_x^2 = P_x P_x = \left(-i\hbar\frac{\partial}{\partial x}\right)\left(-i\hbar\frac{\partial}{\partial x}\right) = -\hbar^2\frac{\partial^2}{\partial x^2}$$

the hamiltonian operator now becomes

$$\hat{H} = \frac{-\hbar^2}{2m}\frac{\partial^2}{\partial x^2}$$

In fact, since we are dealing in one dimension only, we can write this as

$$\hat{H} = \frac{-\hbar^2}{2m}\frac{\mathrm{d}^2}{\mathrm{d}x^2} \tag{2.3}$$

Further examples are given in the problems at the end of the chapter.

2.1.2 *The wavefunction, ψ*

When solving the Schrödinger equation we must find a function of all the particles, ψ, such that Equation (2.1) is obeyed. This type of problem is called an *eigenvalue* problem and the solution ψ is called the *eigenfunction.*

ψ has the property that when it is operated on by $\hat{H}$ the result is to obtain ψ back again—apart from a simple multiplicative factor, E (the eigenvalue).

Finding the eigenfunctions is the most difficult part of solving quantum mechanical problems and we shall see that in many cases it is better to construct simple models, with readily obtained solutions, rather than trying to solve the real problem directly. Provided the models bear a reasonable resemblance to reality we can use them to explain the observed spectra.

Upon solving the Schrödinger equation the eigenvalue obtained, E, is just the energy of the system—the quantity we require. It is usually found that Equation (2.1) does not yield a unique solution but rather many different ones $\psi_1, \psi_2, \psi_3, \ldots$ with associated energies E_1, E_2, E_3. A simple example of this is found by examining the solutions to Equation (2.3)—for a bounded region these have the form:

$$\psi = \sin nx \tag{2.4}$$

we can confirm this by operating with $\hat{H}$

$$\hat{H}\psi = \frac{-\hbar^2}{2m}\frac{d^2}{dx^2}(\sin nx)$$

$$= \frac{n^2\hbar^2}{2m}\sin nx = \frac{n^2\hbar^2}{2m}\psi \tag{2.5}$$

We can see that Equation (2.1) is fulfilled with $E = (n\hbar)^2/2m$. There are an infinite number of solutions depending on the value of n, each with its own energy value.

The eigenfunction or wavefunction ψ provides the *probability* of finding a particle in a particular region of space. The manner in which it does so is not quite straightforward and it is common to find that elementary books cover this point incorrectly; so it is worthwhile looking at a simple case in more detail—this we will do in Section 2.3. For the moment a simple statement will suffice.

If $\psi(x)$ is a solution of the Schrödinger equation for a single particle confined to one dimension then $|\psi(x)|^2$ is the *probability density* of finding the particle at position x. By probability density we mean here probability per unit length. In a three-dimensional case we would mean probability per unit volume. Returning to the one-dimensional case we must multiply $|\psi(x)|^2$ by a length to obtain a probability.

2.1.3 *The energy, E*

The eigenvalues of Equation (2.1) are the energy levels of the system under study. It is these that we are most interested in and, for spectroscopy at least, very often no explicit knowledge of the wavefunctions themselves is required. In general one finds a large number of discrete values—frequently infinite, as in the case of the hydrogen atom:

$$E_n = -\frac{me^4}{8\varepsilon_0^2 h^2}\frac{1}{n^2} \qquad n = 1, 2, \ldots \tag{2.6}$$

The energy zero in this case is for the electron at an infinite distance from the nucleus, i.e. a free electron. This means that the bound energy

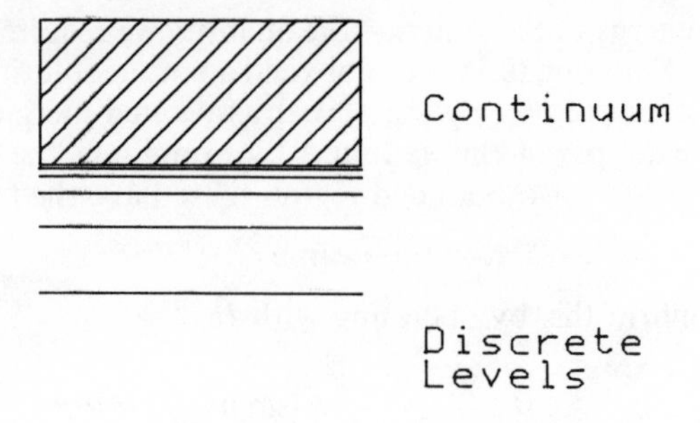

Figure 2.1 Bound and continuum energy levels for the hydrogen atom

levels all have a *negative* sign (lower in energy) when measured with respect to this zero.

It is also possible for some problems to obtain a *continuum* of values for E. In this case the energy is no longer discrete but behaves much more like a continuous classical variable. Such a continuum is obtained for the energy levels of the electron in the hydrogen atom once the energy exceeds the ionization potential of hydrogen. In this case the electron becomes a free particle.

Program 2.1 calculates the first ten bound energy levels of the hydrogen atom using Equation (2.6).

Program 2.1 Hydrogen atom energy levels

```
100  REM  HYDROGEN
110 ME = 9.109E - 31: REM  electron mass kg
120 EC = 1.6021E - 19: REM  electron charge C
130 EO = 8.8542E - 12: REM   permittivity of vacuum F m-1
140 H = 6.62559E - 34: REM  Planck constant J s
150  PRINT "HYDROGEN ATOM LEVELS"
160 A = ME / (8 * H)
170 A = A * EC * EC / H
180 A = A * (EC / EO) * (EC / EO)
190  PRINT
200  PRINT "LEVEL","ENERGY/J"
210  PRINT
220  FOR N = 1 TO 10
230 E =  - A / (N * N)
240  PRINT N,E
250  NEXT
```

```
HYDROGEN ATOM LEVELS

LEVEL           ENERGY/J

1               -2.17967341E-18
2               -5.44918353E-19
3               -2.42185934E-19
4               -1.36229588E-19
5               -8.71869365E-20
6               -6.05464836E-20
7               -4.44831309E-20
8               -3.40573971E-20
9               -2.69095483E-20
10              -2.17967341E-20
```

Program notes

(1) The electron mass and charge are given explicitly in S.I. units, as are the permittivity of free space and Planck's constant.
(2) Because of the very large powers of ten involved in the constants it is necessary (lines 160–180) to be careful in evaluating their products and divisors. Many micros have a limited size for the exponents of real numbers which is not much different from Planck's constant. In normal situations it would be advisable to evaluate the powers of ten by hand and just supply the final number.

2.2 The Born–Oppenheimer approximation

This approximation is fundamental to our treatment of molecular systems and arises from a consideration of the relative motions of the nuclei and the electrons. Nuclei have a much greater mass than electrons (1800 times greater at least) so they move very much more slowly than the electrons. The consequence of this is that the electrons quickly rearrange themselves following a change in the position of the nuclei. In fact the motion of the electrons is so rapid that we can solve the Schrödinger equation for the electrons independently of the one for the nuclei. When solving the electronic Schrödinger equation one treats the nuclei as stationary particles, in this way the nuclear coordinates become *parameters* in the electronic problem. This is called the *Born–Oppenheimer* or *Fixed nucleus* approximation.

In this approach the nuclear kinetic energy term disappears and the nuclear potential energy V_R is a constant (for a given position of the nuclei). We now solve an equation of the form:

$$(\hat{H}_e + \hat{V}_R)\psi_e = U\psi_e \tag{2.7}$$

where $\hat{H}_e$ is a purely *electronic* hamiltonian. Since V_R is a constant the energy will be of the form

$$U = E_e + V_R \tag{2.8}$$

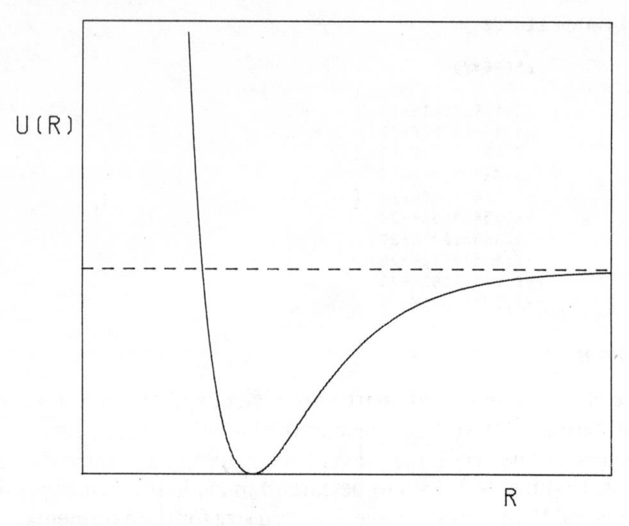

Figure 2.2 A typical ground state potential energy curve for a diatomic molecule

We can solve the *electronic* problem at many different values of R to 'map out' the electronic energy as a function of nuclear coordinates. In the case of a typical diatomic molecule this would resemble the curve in *Figure 2.2.*

The nuclear motions are obtained by solving the *nuclear hamiltonian* problem:

$$\hat{H}_{\mathrm{n}}\psi_n = E_n\psi_n \tag{2.9}$$

When solving the nuclear problem, U (Equation (2.8)) is the potential in which the nuclei move—so, in principle, solutions to the electronic problem are required first. Fortunately, as we shall see later, one often only needs to know the shape of U close to the bottom of the potential well and we can approximate this with a simple function such as a quadratic or Morse curve with reasonable accuracy. This removes the necessity of having to solve the electronic problem explicitly.

As a result of separating out the electronic and nuclear motions the total wavefunction becomes a *product* of the electronic and nuclear wavefunctions.

$$\psi = \psi_{\mathrm{e}}\psi_{\mathrm{n}} \tag{2.10}$$

and the molecular energy is a *sum* of the two contributions.

$$E_{\mathrm{tot}} = E_{\mathrm{e}} + E_{\mathrm{n}} \tag{2.11}$$

It is a general feature of quantum mechanics that if one can separate the hamiltonian operator into independent parts then the total wavefunction will be a product of wavefunctions arising from each part and the total energy will be a simple summation of energies from each part.

This principle is also used to simplify the treatment of vibrating and rotating molecules. In essence, one assumes that the translations, vibrations and rotations of a molecule are independent of each other and the nuclear hamiltonian can now be written as:

$$\hat{H}_n = \hat{H}_t + \hat{H}_v + \hat{H}_r \tag{2.12}$$

with the result that the wavefunction takes the form

$$\psi_n = \psi_t \psi_v \psi_r \tag{2.13}$$

and the energy as

$$E_n = E_t + E_v + E_r \tag{2.14}$$

We may now solve the Schrödinger equation for each type of motion independently. This provides a good 'first order' description of the observed energy levels but it needs modifying if accurate values are required. The approximations involved in separating the vibrations from the rotations are the poorest and vibration–rotation interactions are considered explicitly in Chapter 4.

2.3 Wavefunctions and normalization

A fundamental property of all wavefunctions is that they must have 'quadratic integrability'. This means that

$$\int \psi^* \psi \, d\tau = K \tag{2.15}$$

where $d\tau$ is used to denote a general volume element (all variables) and the integration is over all space. K is a finite constant and ψ^* represents the complex conjugate of ψ (in case the wavefunction is complex).

We are free to scale the wavefunction by any constant factor and we choose one that yields

$$\int \psi^* \psi \, d\tau = 1 \tag{2.16}$$

In this case the wavefunction is said to be *normalized*. In the case of a *single* particle this leads directly to a simple physical interpretation in terms of probability. We say that $\psi^* \psi \, d\tau$ is the probability of finding the particle in a volume element $d\tau$.

Since the particle must be located somewhere, then integrating over all space must yield a probability of one. In many cases the particle is

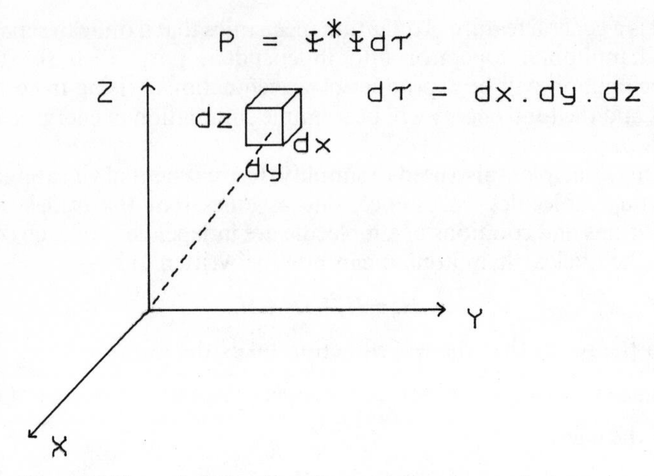

Figure 2.3 The probability of locating an electron in a particular region of space

not free to move over the whole space and the integration is restricted to regions where the wavefunction has non-zero values.

As an example let us normalize the wavefunction (2.4) assuming x can lie in the range 0–π. We write (2.4) as

$$\psi = A \, . \sin nx \tag{2.17}$$

where A is the *normalizing constant*. Equation (2.16) now produces

$$\int_0^{\pi} \psi^* \psi \, dx = \int_0^{\pi} (A \, . \sin nx)^2 \, dx = 1$$

$$A^2 \int_0^{\pi} \sin^2 nx \, dx = 1$$

$$A = \sqrt{\frac{2}{\pi}} \tag{2.18}$$

hence the normalized wavefunction given by

$$= \sqrt{\frac{2}{\pi}} \sin nx \tag{2.19}$$

In the case of many particle wavefunctions the interpretation is a little more complex and the interested reader should consult any introductory text on quantum mechanics for more detail.

2.4 Orthogonality and symmetry

2.4.1 *Orthogonality*

A general property of wavefunctions obtained by solving the Schrödinger equation is that they are orthogonal. This leads to considerable simplification in spectroscopic problems. Orthogonality is defined by

$$\int \psi_m^* \psi_n \, d\tau = 0 \tag{2.20}$$

where ψ_n and ψ_m are wavefunctions corresponding to *different* energies E_n, E_m. If the wavefunctions have the same energy then Equation (2.20) is not necessarily true although we can always construct alternative solutions from them which do obey Equation (2.20).

2.4.2 *Symmetry*

Another fundamental and very useful property of wavefunctions is their symmetry. The use of symmetry in molecular spectroscopy is now firmly established—especially in the treatment of polyatomic systems. For instance, it is often easier to determine selection rules from symmetry considerations than it is to evaluate the detailed integrals required otherwise. Unfortunately the systematic application of symmetry requires complex group theoretical techniques—a description of which would double the length of this book! We shall, however, make use of simple symmetry arguments whenever this aids our understanding.

In essence the symmetry property of a wavefunction is determined by how it behaves upon changing the sign of a coordinate, e.g.

$$\psi(r) = \psi(-r) \qquad \text{symmetric} \quad (+)$$
$$\psi(r) = -\psi(-r) \qquad \text{antisymmetric} \ (-) \tag{2.21}$$

where r is the coordinate. From this definition we can see $\cos(x)$ is symmetric in x, but $\sin(x)$ is antisymmetric. Equally $\cos(x).\sin(y)$ is symmetric in x and antisymmetric in y, and antisymmetric overall in x and y. If a wavefunction is the product of several individual functions we can determine the *overall* symmetry very simply; we assign a + sign to each symmetric function and a − sign to each antisymmetric function, the overall symmetry (sign) is found by multiplying out the signs using the laws of simple arithmetic.

2.4.3 *Averages*

We have seen that to find the energy of a system we need to construct the hamiltonian (energy) operator and use Equation (2.1) to provide the answer. This can be regarded as 'interrogating' the wavefunction with the operator $\hat{H}$. In fact, this is exactly how quantum mechanics calculates physical observables—it interrogates the wavefunction with a suitable operator. To do this we write down the classical expression for the property we require and convert it (using *Table 2.1*) into an operator. We then use an equation analogous to Equation (2.1) to obtain the desired value (eigenvalue).

For instance, if we wanted to know the square of the linear momentum in the x direction we would write

$$P_x^2 = (m\dot{x})^2 \qquad \text{classically}$$

$$\hat{P}_x^2 = -\hbar^2 \frac{\partial^2}{\partial x^2} \qquad \text{operator}$$

and we would have to solve

$$\hat{P}_x^2 \psi = p\psi \qquad (2.22)$$

The number p is the desired value. In many cases it would be found that an equation of the form of Equation (2.22) is not obeyed and a simple number is not produced. This is because quantum mechanics is far less deterministic than classical mechanics and not all classical variables have quantum counterparts. As a result of this limitation, which is closely tied to the Heisenberg Uncertainty Principle, the best we can hope for is an *average* value of the variable concerned. These averages are found in the following way:

$$\langle q \rangle = \frac{\int \psi^* \hat{Q} \psi \, d\tau}{\int \psi^* \psi \, d\tau} \qquad (2.23)$$

where $\hat{Q}$ is the operator and $\langle q \rangle$ its average value. For normalized wavefunctions the denominator is unity and can be omitted.

2.5 The Boltzmann distribution

A spectroscopic measurement is not made on an individual isolated molecule, but rather on a sample containing a very large number of molecules. Each molecule will have associated with it certain amounts of translational, rotational, vibrational and electronic energy. That is, they will occupy a variety of the different energy levels available to the molecule. The measured spectrum will thus contain contributions from a host of different energy states of the molecule. In order to understand the *intensities* of observed spectral lines it is necessary to know how many molecules there are in each energy state. Provided the molecules are in thermal equilibrium with each other, then the answer is given by the *Boltzmann Distribution*. This states that if we have two energy levels E_i and E_j ($E_i > E_j$), then the ratio of the number of molecules in each state is given by:

$$\frac{n_i}{n_j} = \frac{g_i}{g_j} e^{-(E_i - E_j)/kT} \tag{2.24}$$

where T is the absolute temperature, k is the Boltzmann constant and g_i is the *degeneracy* of energy level i.

The degeneracy term arises if more than one wavefunction shares the same energy, an example would be the p orbitals in the hydrogen atom which have a degeneracy of 3 (P_x, P_y, P_z) or the d orbitals with a degeneracy of 5. In fact for the hydrogen atom each quantum number n has n^2 degenerate levels associated with it.

It is more usual to relate the population of molecules in level i to the population in the lowest level.

$$\frac{n_i}{n_0} = \frac{g_i}{g_0} e^{-(E_i - E_0)/kT} \tag{2.25}$$

The total number of molecules, N, is just found by summing all the individual populations.

$$N = n_0 + n_1 + n_2 + \cdots$$

$$= n_0 + n_0 \frac{g_1}{g_0} e^{-(E_1 - E_0)/kT} + n_0 \frac{g_2}{g_0} e^{-(E_2 - E_0)/kT} + \cdots$$

$$N = \frac{n_0}{g_0} \sum_{i=0} g_i e^{-(E_i - E_0)/kT} = \frac{n_0}{g_0} Q \tag{2.26}$$

The summation, Q, is called the *molecular partition function*. It is a pure number whose size depends on the degeneracies and energy level spacings of the molecule under consideration. It is a measure of the number of energy levels which are populated at thermal energy kT. We would thus expect a number close to 1 for non-degenerate electronic energy levels but a much higher value for molecular rotation where the energy spacings are much less. Typical values for Q are: electronic 1; vibrational 1–10; rotational 10–100 and translational $\sim 10^{30}$.

Using Equation (2.26) we can write Equation (2.25) as

$$n_i = \frac{N}{Q} g_i e^{-(E_i - E_0)/kT} \tag{2.27}$$

which yields the number of molecules in each energy level. In fact, if relative intensities are required, it is often more convenient to use Equation (2.25). It can be seen that in the case of non-degenerate levels the populations fall off exponentially with energy. Program 2.2 calculates the relative populations for the first ten energy levels assuming non-degenerate levels with linear energy spacings of 0.1 kT, kT and 10 kT respectively.

Program 2.2 Relative Boltzmann populations

```
100  REM  BOLTZMANN
110 E1 = 0.1: REM  0.1 kT
120 E2 = 1: REM  kT
130 E3 = 10: REM  10 kT
140  PRINT "RELATIVE POPULATIONS"
150  PRINT
160  PRINT "I    0.1kT","     kT","     10kT"
170  FOR I = 0 TO 10
180 N1 =  EXP ( - E1 * I)
190 N2 =  EXP ( - E2 * I)
200 N3 =  EXP ( - E3 * I)
210  PRINT I;"  ";N1,N2,N3
220  NEXT
230  END
```

RELATIVE POPULATIONS

I	0.1kT	kT	10kT
0	1	1	1
1	.904837418	.367879441	4.53999298E-05
2	.818730753	.135335283	2.06115362E-09
3	.740818221	.0497870684	9.35762298E-14
4	.670320046	.0183156389	4.24835426E-18
5	.60653066	6.737947E-03	1.92874985E-22
6	.548811636	2.47875218E-03	8.75651077E-27
7	.496585304	9.11881966E-04	3.97544975E-31
8	.449328964	3.35462628E-04	1.8048514E-35
9	.40656966	1.23409804E-04	0
10	.367879441	4.53999298E-05	0

2.6 Light absorption, selection rules and nomenclature

2.6.1 *Light absorption—a physical picture*

The detailed quantum description of the interaction of matter with light is beyond the scope of this book. The interested reader is referred to References 4 and 5 for more comprehensive treatments. We must be content here with understanding the basic physical principles and utilizing the important results without proof.

Light consists of oscillating electric and magnetic fields propagating through space, as shown schematically in *Figure 2.4* for plane

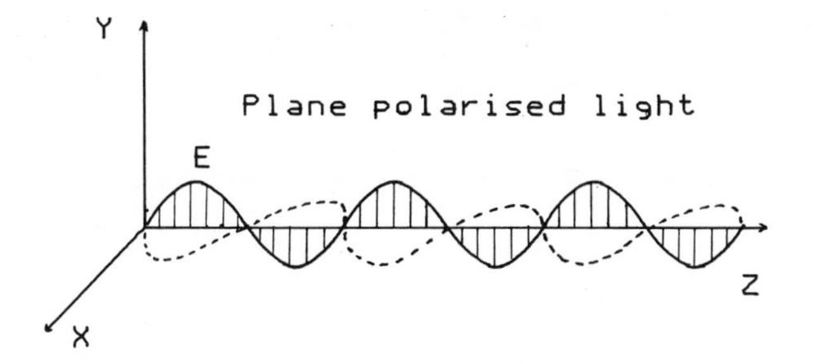

Figure 2.4 Plane polarized light: the electric field is in solid, the magnetic field in dashed

polarized light. In unpolarized light the electric and magnetic fields are not confined to a single plane—although they are always perpendicular to each other at any one moment.

Molecules interact with the light via either the changing electric or magnetic field. The topics in this book arise exclusively from the electric field component, which is not to say that the magnetic field is unimportant (it is responsible for nuclear magnetic resonance spectroscopy, for instance), but rather that the electric field interactions dominate the types of transitions studied here.

We can describe the electric field by

$$E = E^0 \cos 2\pi\nu t \tag{2.28}$$

where ν is the frequency of the light. To a good first approximation a molecule couples to this electric field only if it possesses a fluctuating electric dipole moment, or more accurately, a fluctuating component of the electric dipole moment vector.

Since in this book we are concerned with spectroscopic transitions arising from molecular rotations, vibrations and changes of electronic state, we must now consider how these motions produce such time-dependent dipole moments.

The dipole moment of a series of charged particles is defined by

$$\vec{\mu} = \sum q_i \vec{r}_i \tag{2.29}$$

where q_i is the charge on particle i and $\vec{r}_i$ the vector defining its position. If a molecule has overall neutrality (i.e. it isn't an ion) then the value of $\vec{\mu}$ is independent of the origin chosen.

A heteronuclear diatomic molecule will exhibit a dipole moment since the electronic charge distribution around each nucleus is not identical. *Figure 2.5* shows how, because of this, a rotating heteronuclear diatomic molecule such as HCl will produce a sinusoidally varying component of the electric dipole moment vector.

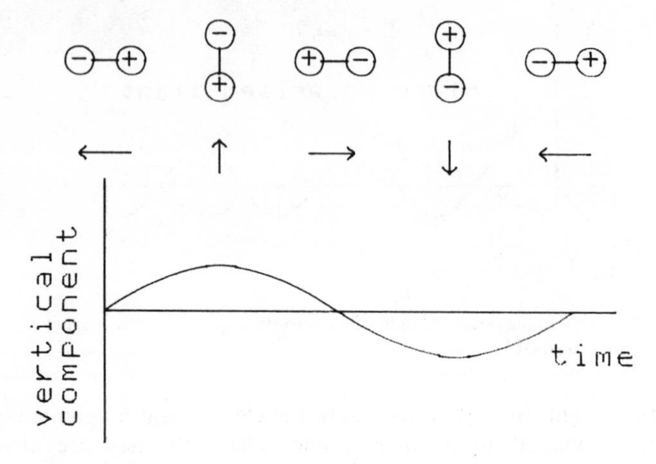

Figure 2.5 The oscillating dipole arising from rotation

We can see from this that a molecule must possess a permanent dipole moment before it can exhibit a rotational spectrum. Molecules such as Cl_2 and H_2 thus have no pure rotational (microwave) spectra.

When a molecule with a permanent dipole moment vibrates, the changing bond length produces a changing dipole moment. However, unlike rotation, molecules with no net dipole can still have fluctuating dipoles associated with vibration. An example can be found by examining the bond stretching motions of CS_2. These consist of a symmetric stretch and an asymmetric stretch. The former has zero dipole moment always, but the latter, which compresses one bond and extends the other, only has an *average* value of zero. This is shown in *Figure 2.6*. It is often relatively easy to decide whether a vibration will be observable spectroscopically by simple considerations of this type.

Unfortunately determining the features that decide whether an *electronic* transition will occur are much more difficult and no simple picture is adequate. We require to evaluate Equation (2.29) for the electrons and this in turn requires a knowledge of the wavefunctions involved in the initial and final states. We shall return to this in Chapter 6.

2.6.2 *Light absorption—a more quantitative treatment*

One of the major consequences of quantizing the energy of a molecule into discrete levels is that only certain frequencies of light can be absorbed or emitted by the molecule. The relationship governing the

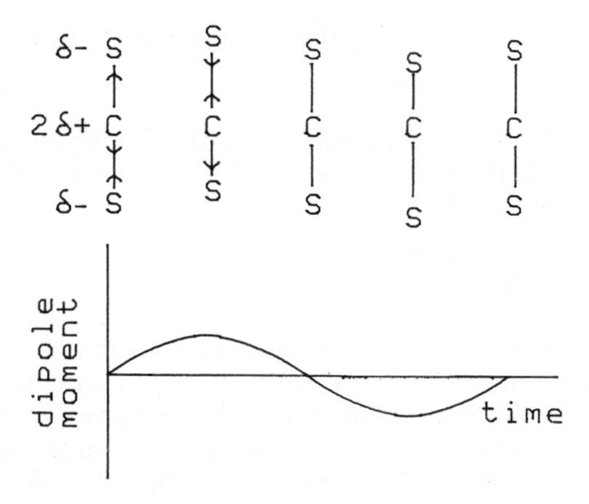

Figure 2.6 The oscillating dipole arising from an asymmetric stretch

energy gap, ΔE, between two levels and the frequency of light, v, which is absorbed or emitted is given by the Bohr–Einstein law:

$$\Delta E = hv \tag{2.30}$$

Classically the energy of interaction between an electric field and a dipole moment is given by

$$E = -\vec{\mu}.\vec{\varepsilon} \tag{2.31}$$

When this term is added to our initial hamiltonian, and time-dependent perturbation theory is applied to the transition between state ψ_m and ψ_n we find that we need to evaluate integrals of the type

$$R_x^{nm} = \int \psi_n^* \hat{M}_x \psi_m \, d\tau \tag{2.32}$$

where

$$\hat{M}_x = \sum e_i x_i \tag{2.33}$$

It can be seen that Equation (2.33) is just the x component of the dipole moment defined in Equation (2.29), where e is the charge on the electron. The integral (2.32) is called a *transition moment* integral and there is one for each of the x, y and z directions. They are important because it is found that the probability of undergoing a transition from ψ_m to ψ_n is proportional to the square of the transition moment integral: $|R^{nm}|^2$.

The intensity of an *emission* line I_{em}^{nm} is defined as: 'the energy emitted by the source per second'. If there are N_n atoms/molecules in the initial state ψ_n, and A_{nm} is the fraction of atoms in this state carrying out transitions to the state ψ_m per second, then

$$I_{\text{em}}^{nm} = N_n h\nu_{nm} A_{nm} \tag{2.34}$$

where $h\nu_{nm}$ is the energy of the photon emitted during the transition. The factor A_{nm} is the *Einstein transition probability for spontaneous emission* and has the form

$$A_{nm} = \left(\frac{1}{4\pi\varepsilon_0}\right)\frac{64\pi^4\nu_{nm}^3}{3hc^3}|R^{nm}|^2 \quad \text{s}^{-1} \tag{2.35}$$

If we consider *absorption* from state ψ_m to ψ_n we must know several additional quantities. These are the number of molecules per unit volume N_m; the number of photons with frequency near ν_{nm} (obtained from the *photon density* $\rho(\nu_{nm})$) which are incident on the sample per unit time; and finally the thickness of the sample Δx. We then have

$$I_{\text{abs}}^{nm} = N_m B_{nm} h\nu_{nm} \rho(\nu_{nm})\,\Delta x \tag{2.36}$$

where B_{nm} is the *Einstein transition probability for absorption* and is given by

$$B_{nm} = \left(\frac{1}{4\pi\varepsilon_0}\right)\frac{8\pi^3}{3h^2}|R^{nm}|^2 \quad \text{m kg}^{-1} \tag{2.37}$$

2.6.3 *Selection rules*

It can be seen from the preceding section that a non-zero value for the transition moment integral is required for a transition to occur. It is often possible to deduce by simple argument whether a transition moment integral will be zero or not without recourse to detailed calculation. This can be on the basis of symmetry considerations or general conservation laws.

A transition is said to be *forbidden* if $|R^{nm}|$ is zero, and *allowed* if it is not. The term 'forbidden' is not as absolute as it may sound and it is often observed that 'forbidden transitions' do in fact occur—but weakly. In some cases these arise from allowed *magnetic dipole* or *electric quadrupole* transitions which are inherently much weaker than electric dipole transitions (about 10^{-4} and 10^{-6} times respectively). In other cases the original treatment of the molecular motions may have been too simple, leading to rules that are inappropriate, as we shall see when treating a vibrating diatomic molecule as a simple harmonic oscillator.

2.6.4 *Experimental quantities*

The absorption of monochromatic light is described experimentally by the Beer–Lambert law

$$I(\nu)=I_0(\nu)\,e^{-\alpha Cl} \tag{2.38}$$

where I and I_0 are the transmitted and incident light intensity respectively, C is the concentration of molecules and l is the pathlength of light through the sample. The constant α is called an *absorption coefficient* and is a function of wavelength. There are two commonly used sets of units for Equation (2.38)

C	p	atm	n	molecules cm^{-3}
l	l	cm	l	cm
α	k	$atm^{-1}\,cm^{-1}$	σ	cm^{-2} $molecule^{-1}$
I	I	W cm^{-2}	I	photons $cm^{-2}\,s^{-1}$

In addition, Equation (2.38) is frequently used with base 10

$$I(\nu)=I_0(\nu)10^{-\varepsilon Cl} \tag{2.39}$$

with units of ε ($dm^3\,mol^{-1}\,cm^{-1}$), l (cm) and C ($mol\,dm^{-3}$). In this case ε is called the *molar absorption coefficient* or *molar extinction coefficient*. We can relate σ, k and ε by

$$\sigma=3.72\times 10^{-20}\,k \qquad (0°C)$$

$$\varepsilon=9.73\,k \qquad (0°C)$$

$$\sigma=3.82\times 10^{-21}\,\varepsilon$$

Real molecules do not absorb at a single wavelength but rather over a band of wavelengths. The total absorption of the band is found by integrating the absorption coefficient across the band to yield the *total absorption coefficient, A*

$$A=\int_{\text{over band}} \varepsilon(\nu)\,d\nu \tag{2.40}$$

A detailed analysis shows that we can relate the total absorption coefficient A to the Einstein coefficient B_{nm} by the expression

$$B_{nm}=\left(\frac{2.303c}{10Lh\nu_{nm}}\right)A \tag{2.41}$$

where L is Avogadro's number, c the velocity of light and A is in units of mol dm^{-3} cm^{-1} s^{-1}, i.e. the integrated molar absorption coefficient.

2.6.5 *Nomenclature*

It is important to adopt a consistent standard for labelling molecular transitions since many different formats have grown up as spectroscopy developed. The system used throughout this book will be the one laid down by the International Joint Commission for Spectroscopy.

All transitions are written with their *upper* state first, regardless of whether they are in absorption or emission. The type of transition is denoted by the direction of an arrow.

$$B \leftarrow A \qquad \text{absorption}$$

$$B \rightarrow A \qquad \text{emission}$$

If primes and double primes are used to distinguish different states then a single prime, ′ denotes the *upper* state and a double prime, ″ the lower state.

$$v' \leftarrow v'' \qquad \text{absorption}$$

2.7 Spectroscopic units

In an ideal world the S.I. system of units would be adopted universally, in which case energy would be in joules or kJ mol^{-1}), frequency in Hz and wavelength in metres. In reality, these units often produce cumbersome numbers involving large powers of ten. As a result of this, and for various historical reasons, a 'witch's brew' of alternative units is widely used in spectroscopy. These tend to be of convenient size for a particular region of the spectrum and are now firmly entrenched in the literature.

An additional complication is that most books on spectroscopy (even some relatively recent ones) use the c.g.s. system and associated Gaussian units for electrostatics and electromagnetism. For the unwary they present enormous scope for error. Special care must be taken when using formulae expressed in non-S.I. units since they often have 'missing' or 'extra' fundamental constants such as h, c and ε_0 depending on the units.

This book will adopt S.I. units except where current practice is overwhelmingly in favour of a non-S.I. unit. All fundamental constants will be in S.I. units in *all* formulae—any compensating numerical factors for non-S.I. values will appear explicitly in the formulae.

Spectroscopy is concerned with the energy *difference* between two states and the absolute values of the energy levels is of little importance. Further, since this energy difference is related to the light frequency by Equation (2.30), it is common to refer to the energy difference by the frequency or wavelength of the light involved.

2.7.1 *Frequency and wavelength*

The frequency of light is expressed in Hz (s^{-1}) or multiples of this such as MHz (10^6 Hz) or GHz (10^9 Hz). The wavelength of light is commonly expressed in angstroms, Å (10^{-10} m); millimicrons, mμ or nanometres, nm (10^{-9} m); microns, μ (10^{-6} m) as well as the more common units of metres and cm.

A particularly common unit is the *wavenumber* $\bar{\nu}$, defined by

$$\bar{\nu}=\frac{1}{\lambda}=\frac{\nu}{c} \quad \text{m}^{-1} \tag{2.42}$$

In fact wavenumbers are almost universally given in cm^{-1} and we should write Equation (2.42) as

$$\bar{\nu}=\frac{\nu}{100c} \quad \text{cm}^{-1} \tag{2.43}$$

Figure 2.7 shows schematically the region of the electromagnetic spectrum dealt with in this book.

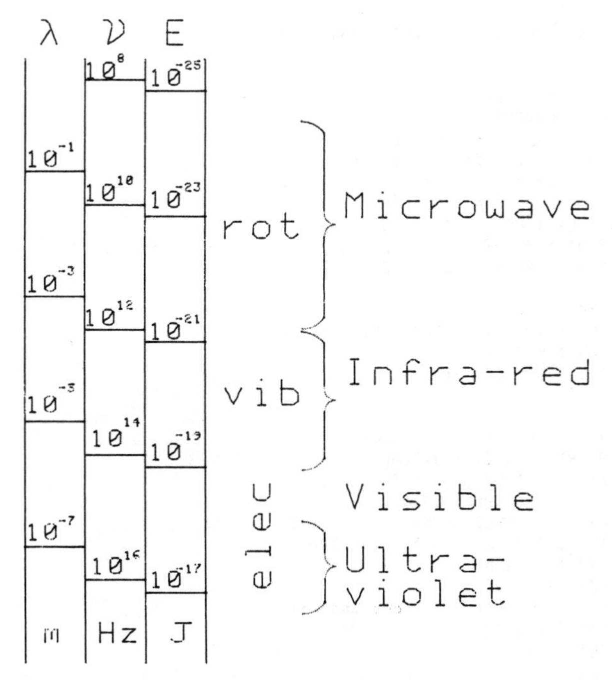

Figure 2.7 The region of the electromagnetic spectrum dealt with in this book

2.7.2 *Energy*

The S.I. unit of energy, the joule (molecule^{-1}), is not widely used and a more common acceptable unit is the kJ mol^{-1}. However, several other units are in common use. These arise from the fact that the frequency of the light in a transition is directly proportional to the energy gap, hence any frequency unit can be used as an energy unit provided one remembers the proportionality constant.

This is an unfortunate practice since considerable confusion can arise as to whether the unit is being used as a frequency or energy unit. For instance, 1 MHz represents 10^6 Hz or $10^6 \times h$ joules. Wavenumbers can also serve as energy units and they are converted to joules by Equation (2.44)

$$E = 100hc\bar{v} \tag{2.44}$$

where $\bar{v}$ is in cm^{-1}.

A final energy unit is the *electron volt*, eV. This is the kinetic energy gained by an electron accelerated through a potential difference of 1 V. It is commonly used in the ultra-violet region of the spectrum.

Program 2.3 Energy unit conversions

```
90  DIM CF(5),UN$(5)
100  REM  CONVERT
110  GOSUB 2000: REM  Conversion factors
120  GOSUB 1000: REM  Print menu
130  GOSUB 1500: REM  Get units and value
140  IF IU = 0 THEN  GOTO 180
150 EO = EI * CF(IU) / CF(OU)
160  PRINT : PRINT EI;" ";UN$(IU);"  =  ";EO;" ";UN$(OU)
170  PRINT : PRINT : GOTO 130
180  END
1000  REM  Menu
1010  PRINT : PRINT "MENU CHOICES": PRINT
1020  FOR I = 0 TO 5
1030  PRINT I;"  ";UN$(I)
1040  NEXT
1050  RETURN
1500  REM  Get units and value
1510  PRINT : INPUT "INITIAL UNITS (0-5) ? ";IU
1520  IF IU < 0 OR IU > 5 THEN  GOTO 1510
1530  IF IU = 0 THEN  GOTO 1570
1540  PRINT : INPUT "FINAL UNITS (1-5) ? ";OU
1550  IF OU < 1 OR OU > 5 GOTO 1540
1560  PRINT : INPUT "VALUE = ";EI
1570  RETURN
2000  REM  Conversion factors and units
2010  FOR J = 1 TO 5
2020  READ CF(J)
2030  NEXT
2040  FOR J = 0 TO 5
2050  READ UN$(J)
```

```
2060  NEXT
2070  DATA  1,1.66044E-21,6.626E-34,1.986E-23,1.602E-19
2080  DATA  "QUIT","J","kJ mol-1","Hz","cm-1","eV"
2090  RETURN
```

```
MENU CHOICES

0  QUIT
1  J.
2  kJ mol-1
3  Hz
4  cm-1
5  eV

INITIAL UNITS (0-5) ? 5

FINAL UNITS (1-5) ? 2

VALUE = 1

1 eV  =  96.4804509 kJ mol-1


INITIAL UNITS (0-5) ? 4

FINAL UNITS (1-5) ? 3

VALUE = 2503

2503 cm-1  =  7.50220043E+13 Hz


INITIAL UNITS (0-5) ? 0
```

Appendix 1 gives the values of some important fundamental constants and also provides a table of conversion factors between common energy units.

2.8 References

1. Hanna, M.W., *Quantum Mechanics in Chemistry*, W.A. Benjamin Inc. (1969).
2. Atkins, P.W., *Molecular Quantum Mechanics*, Oxford University Press (1970).
3. Levine, I.N., *Quantum Chemistry*, Allyn and Bacon, Boston (1974).
4. Barrow, G.M., *Introduction to Molecular Spectroscopy*, McGraw-Hill (1962).
5. Steinfeld, J.I., *Molecules and Radiation*, The MIT Press (1974).

PROBLEMS

(2.1) Construct the hamiltonian operator corresponding to the classical kinetic energy of a free particle,

$$T = 0.5\ \mathrm{m}(V_x^2 + V_y^2 + V_z^2).$$

(2.2) Show that $\psi = e^{-ax}$ are also eigenfunctions of Equation (2.3). What are their energies?
(2.3) Normalize the wavefunctions $\psi = e^{im\phi}$ where ϕ can take values in the range $0 \leqslant \phi \leqslant 2\pi$ and m is an integer.
(2.4) Modify Program 2.1 to print out the first 10 energy levels of the hydrogen atom in eV, and also the degeneracy of each level.
(2.5) Draw the energy levels from Problem (2.4) to scale and label them with the standard nomenclature (1 s, 2 s, ...).
(2.6) Modify the program from Problem (2.4) to calculate the energy *difference* between adjacent pairs of energy levels.
(2.7) Write a program to test if a function $f(x)$ is symmetric, antisymmetric or neither, with respect to x. (Hint: evaluate $f(x)$ and $f(-x)$ at sample points.)
(2.8) Do you think the method suggested for Problem (2.7) is foolproof?
(2.9) For the normalized wavefunction $\psi = (2/a)^{\frac{1}{2}} \sin(n\pi x/a)$, calculate the *average* value expected for the linear momentum P_x if x is restricted to the range $0 \leqslant x \leqslant a$. (Hint: $\int \sin(mx) . \cos(mx) . \mathrm{d}x = (1/2m) \sin^2(mx)$.)
2.10) Modify Program 2.2 to incorporate degeneracy factors such that each level J is $(2J+1)$ degenerate.
(2.11) If a transition from the ground state to the first excited state requires a photon of frequency 600 GHz, what will be the equilibrium ratio of the population in the first excited state to the ground state at (a) 30 K, (b) 300 K, and (c) 3000 K?
(2.12) Which of the following molecules will possess a rotational (microwave) spectrum; N_2, IBr, CO_2?
(2.13) Which of the molecules in Problem (2.4) will possess a vibrational (infra-red) spectrum?
(2.14) Write a program that reads in the energy of a transition in cm^{-1} and prints out its energy in J, kJ mol^{-1}, eV and also provides the frequency and wavelength of the photon in Hz and metres.

Chapter 3

Rotational spectra

ESSENTIAL THEORY

We now come to the treatment of molecular spectra and demonstrate how the quantum description of molecules both explains the observed spectra and provides quantitative structural information. In this chapter we look at the absorption of radiation in the microwave region of the spectrum corresponding to the excitation of rotational motions of molecules. No allowance will be made here for the fact that molecules also undergo simultaneous vibrations—the so-called *rigid-rotor* approximation. In the next chapter the coupling of rotation and vibration will be examined explicitly.

A good starting point is the classical rotation of a linear system of masses joined by weightless rods (bonds), the simplest case being the diatomic molecule.

3.1 Diatomic molecules

It was shown in Section 2.6, p. 28 that diatomic molecules with a permanent dipole moment can absorb electromagnetic radiation via their rotational motion. Thus CO, NO and HCl will all exhibit rotational spectra but S_2 or H_2 will not. A necessary prerequisite to the quantum treatment of the motion is a classical description of the rotational energy.

3.1.1 *Diatomics—classical rotation*

The rotation of a diatomic molecule is best described in terms of its angular velocity, ω, about the centre of gravity of the molecule.

$$\omega = \frac{V_1}{r_1} = \frac{V_2}{r_2} \qquad \text{rad s}^{-1} \tag{3.1}$$

where V_i is the linear velocity and r_i the distance of particle i from the centre of gravity. Since there are 2π radians in a circle then the number of revolutions per second is $\omega/2\pi$.

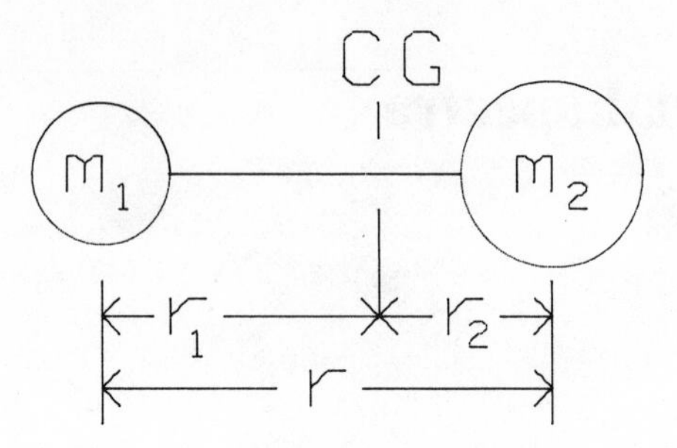

Figure 3.1 A diatomic molecule

An important quantity for describing the energy of rotation is the moment of inertia, I, defined by

$$I=\sum_i m_i r_i^2 \tag{3.2}$$

we can obtain expressions for the r_i from the definition of the centre of gravity

$$m_1 r_1 = m_2 r_2 \tag{3.3}$$

and the fact that the bond length is given by $r = r_1 + r_2$. We obtain

$$r_1=\left(\frac{m_2}{m_1+m_2}\right) r \tag{3.4}$$

and

$$r_2=\left(\frac{m_1}{m_1+m_2}\right) r$$

Applying Equation (3.4) to Equation (3.2) we find

$$I=\left(\frac{m_1 m_2}{m_1+m_2}\right) r^2 = \mu r^2 \tag{3.5}$$

where μ is the reduced mass of the system

$$\mu=\left(\frac{m_1 m_2}{m_1+m_2}\right) \tag{3.6}$$

The kinetic energy, T, of the rotating molecule is now easily calculated from

$$\begin{aligned} T &= \tfrac{1}{2}\sum m_i v_i^2 = \tfrac{1}{2}\sum m_i \omega^2 r_i^2 \\ &= \frac{\omega^2}{2}\sum m_i r_i^2 \\ &= \tfrac{1}{2} I \omega^2 \end{aligned} \tag{3.7}$$

In fact, Equation (3.7) is true for any linear molecule, not just diatomics.

3.1.2 *Quantum solution—the rigid rotor*

The Schrödinger equation for the rigid rotor can now easily be set up from Equation (3.7) and the rules given in *Table 2.1*. The potential energy may be set to zero since there is no change in bond length during the rotation. The detailed solution will not be given here but it can be found in any standard text on quantum mechanics. The energy levels obtained are given by the formula

$$E_J = J(J+1)\frac{h^2}{8\pi^2 I} \qquad J = 0, 1, 2\ldots \tag{3.8}$$

The energy in Equation (3.8) is in joules but spectroscopically rotational energies are most commonly expressed in cm^{-1}, we can convert Equation (3.8) to wavenumbers simply by dividing by $100hc$.

$$\bar{E}_J = J(J+1)\frac{h}{800\pi^2 cI} \qquad cm^{-1} \tag{3.9}$$

where the bar is used to denote wavenumbers and the factor of 100 arises from the conversion of m^{-1} to cm^{-1}. The constant factors in Equations (3.8) and (3.9) occur so frequently that they are given the symbols B and $\bar{B}$ respectively

$$B = \frac{h^2}{8\pi^2 I} \text{ joules,} \qquad \bar{B} = \frac{h}{800\pi^2 cI} \quad cm^{-1} \tag{3.10}$$

and we write Equation (3.8) as

$$E_J = BJ(J+1) \tag{3.11}$$

An understanding of the physical constraints that lead to Equation (3.11) can be obtained by imposing quantization of angular momentum on the classical problem, just as Bohr did in his original treatment of the hydrogen atom. Classically the angular momentum of a rotating system is defined as

$$\begin{aligned} L &= \sum m_i v_i r_i \\ &= \sum m_i r_i^2 \left(\frac{v_i}{r_i}\right) = I\omega \end{aligned} \tag{3.12}$$

if we impose the Bohr quantization condition on the angular momentum we have

$$I\omega = J\frac{h}{2\pi} \qquad J = 0, 1, 2\ldots \tag{3.13}$$

furthermore, we can write Equation (3.7) as

$$T = \frac{1}{2I}(I\omega)^2 = \frac{1}{2I}\left(\frac{h}{2\pi}J\right)^2$$

$$T = \frac{h^2}{8\pi^2 I}J^2 = BJ^2 \qquad (3.14)$$

We can see that this is very similar to the correct quantum solution (3.11), except that J^2 must be replaced by $J(J+1)$. The quantization of the rotational energy is thus a direct consequence of the quantization of the angular momentum of the rotating molecule.

3.1.3 *Energy levels and degeneracy*

The energy level values given by Equation (3.11) can be seen to increase rapidly with J, that is, the levels diverge. This is shown schematically in *Figure 3.2.*

Program 3.1 calculates the moment of inertia and the rotational energy levels for diatomic molecules. It requires only the masses and bond lengths to do this. It is good practice to ask for quantities in convenient units and to convert these internally to S.I. quantities. For this reason the program inputs bond lengths in Angstroms and

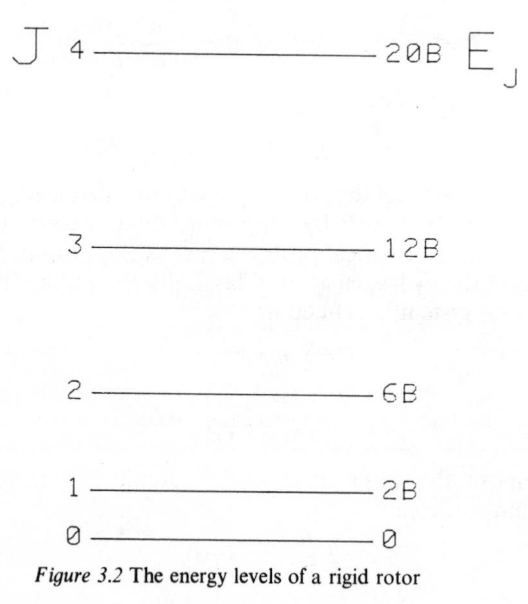

Figure 3.2 The energy levels of a rigid rotor

masses in amu since these represent the most common units for these quantities.

Program 3.1 ROTLEV: Diatomic rotational energy levels

```
100  REM  ROTATIONAL ENERGY LEVELS
110  GOSUB 1000: REM  GET DATA
120 MU = M1 * M2 / (M1 + M2)
130 I = MU * R * R: REM  MOMENT OF INERTIA
140 ZN = I:ZP = 3: GOSUB 5000:I$ = ZN$: REM  FORMAT NUMBER
150  PRINT : PRINT "MOMENT OF INERTIA = ";I$;" Kg m2 mol-1"
160  PRINT : PRINT "J      E/J         E/cm-1": PRINT
170 B = ((H * L / I) * H) / (8 * PI * PI)
180 BW = B / (H * C * 100): REM  cm-1
190  FOR J = 0 TO 9
200 EJ = J * (J + 1) * B:EW = J * (J + 1) * BW
210 ZN = EJ:ZP = 3: GOSUB 5000:EJ$ = ZN$
220 ZN = EW: GOSUB 5000:EW$ = ZN$
230  PRINT J;"   ";EJ$,EW$
240  NEXT
250  END
1000  REM  DATA INPUT
1010  PRINT : PRINT "DIATOMIC DATA": PRINT
1020  PRINT : INPUT "MASS 1 (AMU) = ";M1
1030  PRINT : INPUT "MASS 2 (AMU) = ";M2
1040  PRINT : INPUT "BOND LENGTH (ANGSTROMS) = ";R
1050 M1 = M1 / 1000:M2 = M2 / 1000: REM  Kg mol-1
1060 R = R * 1E - 10: REM  m
1070 H = 6.6256E - 34: REM  Plank's constant
1080 L = 6.0225E23: REM  Avagadro's number
1090 C = 2.9979E8: REM  Vel of light
1100 PI = 4 *  ATN (1)
1110  RETURN
5000  REM  NUMBER FORMATER
5010  REM  ZP = NUMBER OF PLACES IN MANTISSA
5020  REM  ZN = NUMBER
5030  IF ZN = 0 THEN ZN$ = "0": GOTO 5090
5040 ZE =  INT ( LOG (ZN) /  LOG (10)): REM  EXPONENT
5050 ZM = ZN / (10 ^ ZE): REM  MANTISSA
5060 ZR =  INT (ZM * (10 ^ ZP) + 0.5) / (10 ^ ZP)
5070 ZN$ =  STR$ (ZR) + "          "
5080 ZN$ =  LEFT$ (ZN$,ZP + 2) + " E" +  STR$ (ZE)
5090  RETURN
```

```
DIATOMIC DATA


MASS 1 (AMU) = 1

MASS 2 (AMU) = 35

BOND LENGTH (ANGSTROMS) = 1.275

MOMENT OF INERTIA = 1.58  E-23 Kg m2 mol-1

J     E/J          E/cm-1
```

```
0  0            0
1  4.237 E-22   2.133 E1
2  1.271 E-21   6.4   E1
3  2.542 E-21   1.28  E2
4  4.237 E-21   2.133 E2
5  6.356 E-21   3.2   E2
6  8.898 E-21   4.48  E2
7  1.186 E-20   5.973 E2
8  1.525 E-20   7.68  E2
9  1.907 E-20   9.6   E2

]


DIATOMIC DATA


MASS 1 (AMU) = 127

MASS 2 (AMU) = 127

BOND LENGTH (ANGSTROMS) = 2.666

MOMENT OF INERTIA = 4.513 E-21 Kg m2 mol-1

J      E/J          E/cm-1

0  0            0
1  1.484 E-24   7.47  E-2
2  4.451 E-24   2.241 E-1
3  8.903 E-24   4.482 E-1
4  1.484 E-23   7.47  E-1
5  2.226 E-23   1.121 E0
6  3.116 E-23   1.569 E0
7  4.155 E-23   2.092 E0
8  5.342 E-23   2.689 E0
9  6.677 E-23   3.362 E0
```

Program notes

(1) The moment of inertia is given in kg m^2 *mol*$^{-1}$ rather than *molecule*$^{-1}$ to avoid underflow errors on most microcomputers. To convert to molecule^{-1} divide by Avogadro's constant.
(2) The program contains a routine to format numbers to a set number of decimal places. It converts a number to a format *n.nnn Emm* where *E* represents the power of 10. This has been used to round the calculations to realistic physical accuracy and should always be done in final versions of programs, although pressure of space means that it has been omitted from most subsequent programs in this book.

A further point that emerges from the quantum solution is that the energy levels are degenerate. That is, for a given energy, E_J, there is more than one wavefunction that will produce that energy. In fact the Jth energy level is found to be $(2J+1)$ degenerate. The degeneracies for $J=0, 1, 2, 3 \ldots$ are thus $1, 3, 5, 7 \ldots$. We shall see that this has a profound effect on the intensity of the lines found in the rotational spectrum.

The degeneracies correspond to certain allowed orientations of the rotation relative to a fixed reference direction, which we are free to call the z direction. We can reveal the original degeneracies by imposing an external field along this axis. This has the effect of lifting the degeneracy so that the different orientations now have slightly different energies. *Figure 3.3* shows schematically what this would look like for $J=1$.

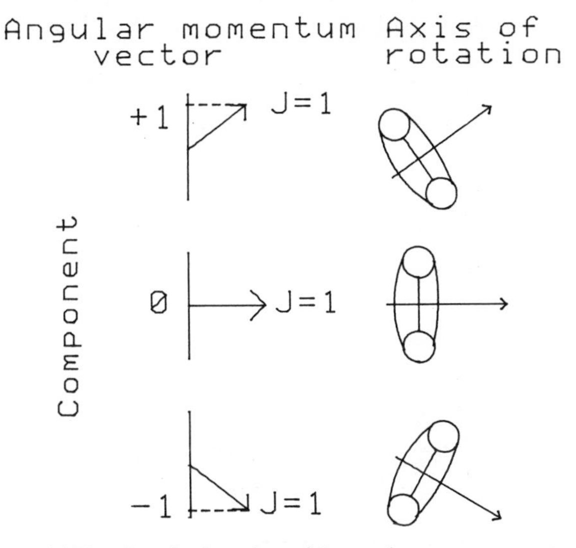

Figure 3.3 The allowed orientations of the angular momentum vector in an external field for $J=1$

For a given value of the rotational quantum number, J, there are components of angular momentum (in units of $h/2\pi$) of $J, J-1, J-2, \ldots 0 \ldots -J$ along the z axis. In our more classical vector picture these correspond to allowed orientations of the angular momentum vector, of length $\sqrt{J(J+1)}h/2\pi$.

3.1.4 *Selection rules*

A rotational spectrum corresponds to *changes* in rotational energy levels. In order to understand the spectrum one must first ascertain which changes are allowed and which are forbidden. We have already stated that a molecule must first possess a permanent dipole moment. A further restriction arises when we consider a detailed analysis of the transition moment integrals (2.32). The mathematical reasoning is too complex to be reproduced here but the result is a simple rule determining the allowed transition:

$$\Delta J = \pm 1 \tag{3.15}$$

$\Delta J = 1$ corresponds to absorption and $\Delta J = -1$ to emission. This rule is very powerful since it simplifies the spectrum considerably—indeed we need only consider adjacent levels when constructing the spectrum. *Figure 3.4* illustrates how application of this rule leads to the spectral lines arising from molecular rotational transitions.

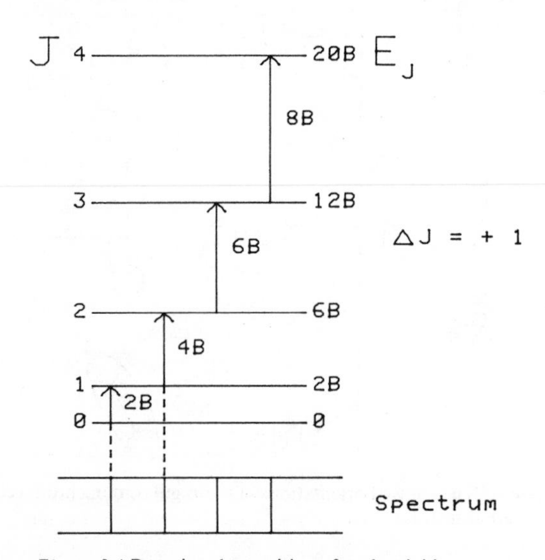

Figure 3.4 Rotational transitions for the rigid rotor

For $\Delta J = 1$ transitions we have

$$E_J = B(J+1)(J+2) - BJ(J+1)$$

$$E_J = 2B(J+1) \tag{3.16}$$

The lines in the rotational spectrum are thus expected to occur at $2B$, $4B$, $6B$.... We see from this that there is a constant spacing of $2B$ between the lines. The measured line spacing is used to evaluate the experimental value of B, which in turn allows us to determine the moment of inertia and hence the diatomic bond length. This is the first example of how molecular spectroscopy can lead directly to structural information.

For instance, the rotational line spacing of $H^{35}Cl$ is found to be 21.18 cm^{-1}. This yields a value for $\bar{B}$ of 10.59 cm^{-1}. From Equation (3.10) we can write

$$I = \frac{h}{800\pi^2 \bar{B}}$$

$$= 2.643 \times 10^{-47} \qquad \text{kg m}^2$$

and from Equation (3.5), remembering to express the masses in kg, we have

$$r^2 = \left(\frac{m_1 + m_2}{m_1 m_2}\right) I \qquad (3.17)$$

$$r = 1.28 \times 10^{-10} \qquad \text{m}$$

We have thus determined the bond length of HCl from a simple spectroscopic measurement. The most accurate determinations of many bond lengths have been obtained from microwave spectroscopy.

3.1.5 *Rotational populations and rotational spectra*

Since almost all rotational spectra are taken in absorption we should address ourselves to how we expect the *intensity* of lines to appear in an absorption spectrum. The selection rule determines which transitions are allowed and Equation (2.36) shows that the intensity of a given absorption line is proportional to the number of molecules in the lower state. We can easily calculate these relative populations from the Boltzmann distribution (2.25), we have

$$N_J = N_0(2J+1)e^{-BJ(J+1)/kT} \qquad (3.18)$$

Figure 3.5 shows the populations obtained for HCl at 300 K.

If we take the other factors in Equation (2.36) to be constant then Equation (3.18) is all we need to describe the relative intensities in the microwave spectrum and *Figure 3.5* is then also a diagram of the rotational spectrum. Assuming, for the moment, that Equation (3.18) is indeed a good representation of the true spectrum we can find the transition with maximum intensity by treating Equation (3.18) as a *continuous* distribution and differentiating with respect to J.

$$\frac{dN_J}{dJ} = e^{-BJ(J+1)/kT}\left\{2 - \frac{B}{kT}(2J+1)^2\right\}$$

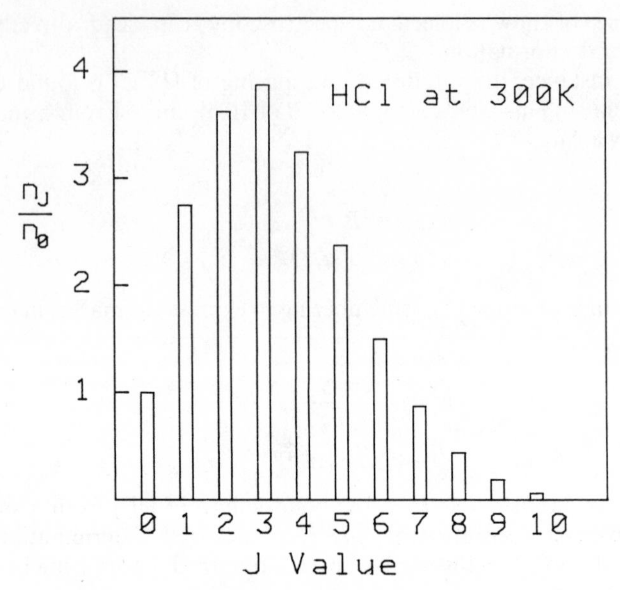

Figure 3.5 The Boltzmann populations for HCl at 300 K

for a maximum $(\mathrm{d}N_J/\mathrm{d}J)=0$ and hence we obtain

$$J_{max}=\sqrt{\frac{kT}{2B}}-\frac{1}{2} \tag{3.19}$$

Since J_{max} must be an integer, one takes the nearest integer value obtained from Equation (3.19) as the maximum.

A more detailed examination, however, reveals that the observed line intensities do not agree with Equations (3.18) and (3.19) exactly and we must re-examine the other terms in Equation (2.36) if we are to be more successful. In general we may still assume that $\rho(\nu_{nm})$ and Δx are constants across the spectrum, but we will have to modify our treatment of the first three terms.

$$I_{abs}^{nm}=N_m B_{mn} h\nu_{nm} \tag{3.20}$$

N_m

For systems with an appreciable excited state population we must replace N_m by the population *difference* between levels *n* and *m*. A more complete analysis indicates that we must replace N_m by

$$\frac{N_m}{g_m}-\frac{N_n}{g_n} \tag{3.21}$$

where g_n and g_m are the degeneracies of each level. Using the Boltzmann expression (2.24) to substitute for N_n, we obtain

$$\frac{N_m}{g_m}\{1-e^{-(E_n-E_m)/kT}\} \quad (3.22)$$

Since we are concerned with energy level J, and we know that $\Delta J = +1$ for absorption, then Equation (3.22) becomes

$$\frac{N_J}{(2J+1)}\{1-e^{-2B(J+1)/kT}\} \quad (3.23)$$

where N_J is given by Equation (3.18) as before.

B_{nm}

The evaluation of the Einstein transition probabilities, B_{nm}, for the rigid rotor is a difficult task and involves considerable mathematical manipulation, however the final result is rather simple:

$$B_{nm} \propto (J+1)\mu^2 \quad (3.24)$$

where μ is the permanent dipole moment.

$h\nu_{nm}$

This term is just the energy of the photon, which is equal to the energy difference between the levels involved. In the case of absorption from a lower level, J, we have from Equation (3.16)

$$h\nu_{nm} = 2B(J+1) \quad (3.25)$$

We can now evaluate Equation (3.20) using Equations (3.23), (3.24), and (3.25)

$$I_{\mathrm{abs}}^{nm} \propto e^{-BJ(J+1)/kT}.(1-e^{-2B(J+1)/kT}).(J+1)^2 \quad (3.26)$$

where only those parts with a J dependency have been retained. We are now in a position to predict the observed rotational spectra of diatomic molecules with greater accuracy than before. All that is required is a knowledge of the masses, bond length and ambient temperature.

Program 3.2 calculates the predicted relative line intensities using Equations (3.18) and (3.26). In both cases the distributions have been normalized to unity at the peak. The program also produces a simple 'bar chart' representation of Equation (3.26).

Program 3.2 ROTSPC: Diatomic rotational spectrum

```
90  DIM ET(50),I1(50),I2(50)
100  REM  ROTATIONAL SPACING
110  GOSUB 1000: REM  GET DATA
120  GOSUB 2000: REM  ROT CONSTANTS
130  GOSUB 3000: REM  INTENSITIES
140  GOSUB 4000: REM  PLOT GRAPH
150  END
1000  REM  DATA INPUT
1010  PRINT : PRINT "DIATOMIC DATA": PRINT
1020  PRINT : INPUT "MASS 1 (AMU) = ";M1
1030  PRINT : INPUT "MASS 2 (AMU) = ";M2
1040  PRINT : INPUT "BOND LENGTH (ANGSTROMS) = ";R
1050  PRINT : INPUT "TEMPERATURE (K) = ";T
1060  PRINT : INPUT "MAXIMUM LEVEL (<50) = ";NL
1070 M1 = M1 / 1000:M2 = M2 / 1000: REM  Kg mol-1
1080 R = R * 1E - 10: REM  m
1090 H = 6.6256E - 34: REM  Plank's constant
1100 L = 6.0225E23: REM  Avagadro's number
1110 C = 2.9979E8: REM  Vel of light
1120 K = 1.3805E - 23: REM  Boltzmann constant
1130 PI = 4 *  ATN (1)
1140  RETURN
2000  REM  ROTATIONAL CONSTANTS
2010 MU = M1 * M2 / (M1 + M2)
2020 I = MU * R * R: REM  MOMENT OF INERTIA
2030 ZN = I:ZP = 3: GOSUB 5000:I$ = ZN$: REM  FORMAT NUMBER
2040  PRINT : PRINT "* * * * * * *": PRINT
2050  PRINT : PRINT "MOMENT OF INERTIA = ";I$;" Kg m2 mol-1"
2060 B = ((H * L / I) * H) / (8 * PI * PI)
2070 ZN = B: GOSUB 5000
2080  PRINT : PRINT "ROTATIONAL CONSTANT B =";ZN$;" J"
2090 BW = B / (H * C * 100): REM  cm-1
2100 ZN = BW: GOSUB 5000
2110  PRINT : PRINT "ROTATIONAL CONSTANT B =";ZN$;" cm-1"
2120  PRINT : PRINT "Abs  Et/cm-1","INTENSITY 1    INTENSITY 2": PRINT
2130 KT = K * T * 5.035E22: REM  kT IN cm-1
2140  RETURN
3000  REM  CALC INTENSITIES
3010  FOR J = 0 TO NL
3020 EW = J * (J + 1) * BW
3030 ET = 2 * BW * (J + 1)
3040 I1(J) = (2 * J + 1) *  EXP ( - EW / KT)
3050 I2(J) = (J + 1) ^ 2 *  EXP ( - EW / KT) * (1 -  EXP ( - ET / KT))
3060 ET(J) =  INT (ET * 100 + .5) / 100: REM  ROUND RESULT
3070  NEXT
3080  REM  NORMALIZE INTENSITIES
3090 M1 = 0:M2 = 0
3100  FOR J = 0 TO NL
3110  IF I1(J) > M1 THEN M1 = I1(J)
3120  IF I2(J) > M2 THEN M2 = I2(J)
3130  NEXT
3140  FOR J = 0 TO NL
3150 I1(J) = I1(J) / M1:I2(J) = I2(J) / M2
3160 I1(J) =  INT (I1(J) * 10000 + 0.5) / 10000
3170 I2(J) =  INT (I2(J) * 10000 + 0.5) / 10000
3180  NEXT
3190  FOR J = 0 TO NL
3200  PRINT (J + 1);"-";J;"  ";ET(J),I1(J),I2(J)
```

```
3210  NEXT
3220  RETURN
4000  REM  PRINT GRAPH
4010  PRINT : PRINT : PRINT "GRAPH OF INTENSITY 2": PRINT
4020  FOR J = 0 TO NL
4030 PS = I2(J) * 30: REM  NUMBER OF STARS
4040 AB$ = "   " +  STR$ (J + 1) + "-" +  STR$ (J) + " "
4050 AB$ =  RIGHT$ (AB$,6): PRINT AB$;
4060  IF PS < 1 THEN  GOTO 4080
4070  FOR P = 1 TO PS: PRINT "*";: NEXT
4080  PRINT : PRINT
4090  NEXT J
4100  RETURN
5000  REM  NUMBER FORMATER
5010  IF ZN = 0 THEN ZN$ = "0": GOTO 5070
5020 ZE =  INT ( LOG (ZN) /  LOG (10)): REM  EXPONENT
5030 ZM = ZN / (10 ^ ZE): REM  MANTISSA
5040 ZR =  INT (ZM * (10 ^ ZP) + 0.5) / (10 ^ ZP)
5050 ZN$ =  STR$ (ZR) + "            "
5060 ZN$ =  LEFT$ (ZN$,ZP + 2) + " E" +  STR$ (ZE)
5070  RETURN
```

```
DIATOMIC DATA


MASS 1 (AMU) = 1

MASS 2 (AMU) = 35

BOND LENGTH (ANGSTROMS) = 1.275

TEMPERATURE (K) = 300

MAXIMUM LEVEL (<50) = 9

* * * * * * *


MOMENT OF INERTIA = 1.58  E-23 Kg m2 mol-1

ROTATIONAL CONSTANT B =2.119 E-22 J

ROTATIONAL CONSTANT B =1.067 E1 cm-1

Abs  Et/cm-1   INTENSITY 1    INTENSITY 2

1-0  21.33     .2639          .027
2-1  42.66     .7148          .1857
3-2  64        .9709          .4863
4-3  85.33     1              .8082
5-4  106.66    .8539          1
6-5  127.99    .6258          .9891
7-6  149.33    .4003          .8124
8-7  170.66    .2257          .5667
9-8  191.99    .1128          .3407
10-9  213.32   .0502          .1783
```

```
GRAPH OF INTENSITY 2

 1-0

 2-1 *****

 3-2 **************

 4-3 ************************

 5-4 ******************************

 6-5 *****************************

 7-6 ************************

 8-7 *****************

 9-8 **********

10-9 *****
```

In fact, due to the difficulty of spanning a large part of the microwave spectrum with one source, experimental microwave spectra do not usually cover more than a few lines of the rotational spectrum. Complete rotational line spectra *are* obtained in the infra-red where they accompany *vibrational* transitions, as we shall see in Chapter 4.

A further careful scrutiny of microwave spectra reveals that the lines are not quite equidistant with a constant spacing of $2B$, but rather they exhibit a spacing that decreases with increasing J. To find the reason for this we must examine the rigid-rotor approximation itself.

3.1.6 *The non-rigid rotor*

A better model of a chemical bond is a spring rather than a rigid rod. The bond is free to stretch and compress with a potential energy similar to that shown in *Figure 2.2*. Masses attached to each other by such a spring will not remain a fixed distance apart as the rotational motion of the system increases, but rather the spring will stretch at higher rotational speeds.

This is exactly what happens to chemical bonds in real molecules as we proceed to higher J levels. This effect can be incorporated into our original hamiltonian to yield the energy levels of the non-rigid rotor, these are found to be given by

$$\bar{E}_J = \bar{B}J(J+1) - \bar{D}J^2(J+1)^2 \qquad \text{cm}^{-1} \tag{3.27}$$

where $\bar{D} = 4\bar{B}^3/\bar{\omega}^2$ and $\bar{\omega}$ is the vibrational frequency.

$\bar{D}$ is known as the *centrifugal distortion constant* and has a value much less than $\bar{B}$. Typical values for HCl are $\bar{B}=10.395\ \mathrm{cm}^{-1}$ and $\bar{D}=0.0004\ \mathrm{cm}^{-1}$.

In absorption, $\Delta J=+1$ and from Equation (3.27) we obtain the energy spacing as

$$\Delta\bar{E}_J=2\bar{B}(J+1)-4\bar{D}(J+1)^3 \tag{3.28}$$

In comparison with Equation (3.16) this yields lines that are not $2\bar{B}$ apart but converge for higher J levels, as expected (see Problem 3.10).

3.2 Linear polyatomic molecules

Linear polyatomic molecules are governed by the same formulae as diatomic molecules and their spectra are interpreted in exactly the same way. However, a single measurement of the line spacing for a polyatomic molecule is not sufficient to determine the bond lengths involved. This is to be expected since Equation (3.10) only relates I and $\bar{B}$. The individual bond lengths cannot be determined from the moment of inertia only.

The Born–Oppenheimer approximation, however, suggests a way around this obstacle. The *electronic* hamiltonian depends only on the nuclear *charge* not the masses, which means that the electronic potential energy curves will be identical for all isotopes of the constituent atoms. We thus expect all isotopic variants of a given molecule to possess identical bond lengths (minima in the electronic potential energy curves). In contrast the moments of inertia depend directly on the masses, as shown in Equation (3.2). As a result of this, isotopic variants of a given molecule will have different moments of inertia and hence different line spacing. We need to measure a rotational line spacing from different isotopes for each bond in the molecule.

For instance to determine the bond lengths in the linear molecule OCS it is sufficient to measure two rotational line spacings such as those from $OC^{32}S$ and $OC^{34}S$. Experimentally we find values of

$$\bar{B}(OC^{32}S)=0.20286\ \mathrm{cm}^{-1}$$

$$\bar{B}(OC^{34}S)=0.19791\ \mathrm{cm}^{-1}$$

The moment of inertia can be calculated in an analogous manner to Section 3.1.1 and it is found to be (see Problem 3.12)

$$I=\frac{m_o m_c r_{co}^2+m_o m_s r_{os}^2+m_c m_s r_{cs}^2}{m_o+m_c+m_s} \tag{3.29}$$

It is easy to obtain values for the moments of inertia from Equation (3.10), which, via Equation (3.29) yield two equations in the two unknowns r_{co} and r_{cs}. Unfortunately, as a little effort will show, the solution of these equations is a far from trivial task. The difficulty lies in the $(r_{co}+r_{cs})^2$ term which introduces cross terms of the form $r_{co}r_{cs}$ into the equations. The equations now become non-linear which is the hardest class of simultaneous equations to solve.

By simple, but somewhat lengthy, substitution we can reduce the problem to a single equation in one of the unknowns. The resulting equation must then be solved numerically. Program 3.3 uses a bisection method to find the solution of this equation. The bisection method is particularly suitable since we know that bond lengths will lie in the range $0.5<r<5.0$ Angstroms. We can thus use these as outer limits, assured that the true solution is to be found between these values. Those readers interested in the technical details, and more sophisticated algorithms are referred to *BASIC Numerical Methods*[1]. For the OCS data we obtain values of

$$r_{co}=1.171\ \text{Å} \tag{3.30}$$

$$r_{cs}=1.552\ \text{Å}$$

In fact, the precision to which most microwave spectra are measured will provide even more significant figures than shown here. If advantage is to be taken of this precision it is necessary to use more accurate values for the atomic masses than the nominal ones given here and to include effects arising from the vibrational motion of the nuclei.

Program 3.3 BONDL: Triatomic bond lengths

```
90  DIM M1(4),M2(4)
100  REM  TRIATOMIC BOND LENGTHS
110  GOSUB 1000: REM  GET DATA
120  GOSUB 2000: REM  ROTATIONAL CONSTANTS
130  GOSUB 4000: REM  BISECTION SOLUTION
140  IF FF = 1 THEN  GOTO 170
150  PRINT : PRINT "R12 = "; INT (R1 * 1000 + .5) / 1000;" ANGSTROMS"
160  PRINT : PRINT "R23 = "; INT (R2 * 1000 + .5) / 1000;" ANGSTROMS"
170  END
1000  REM  DATA INPUT
1010  PRINT : PRINT "LINEAR TRIATOMIC DATA": PRINT
1015  PRINT : INPUT " MOLECULE = ";MN$
1020  PRINT : PRINT "FIRST ISOTOPE DATA"
1030  FOR J = 1 TO 3
1040  PRINT "MASS ";J;: INPUT " (AMU) = ";M1(J)
```

```
1050  NEXT
1060  PRINT : INPUT "ROTATIONAL CONSTANT (cm-1) = ";B1
1070  PRINT : PRINT "SECOND ISOTOPE DATA"
1080  FOR J = 1 TO 3
1090  PRINT "MASS ";J;: INPUT " (AMU) = ";M2(J)
1100  NEXT
1110  PRINT : INPUT "ROTATIONAL CONSTANT (cm-1) = ";B2
1120 H = 6.6256E - 34: REM  Plank's constant
1130 L = 6.0225E23: REM  Avagadro's number
1140 C = 2.9979E8: REM  Vel of light
1150 PI = 4 *  ATN (1)
1160  RETURN
2000  REM  ROTATIONAL CONSTANTS
2010 CF = H * L / (800 * PI * PI * C)
2020 I1 = CF / B1:I2 = CF / B2: REM  I in Kg m2 mol-1
2030  PRINT : PRINT
2040  PRINT "**********": PRINT
2050  PRINT "I1 = ";I1;" Kg m2 mol-1"
2060  PRINT : PRINT "I2 = ";I2;" Kg m2 mol-1"
2070 I1 = I1 * 1E23: REM  I in  amu angstrom2
2080 I2 = I2 * 1E23
2090  REM  SET UP CONSTANT FACTORS
2100 M1(4) = M1(1) + M1(2) + M1(3)
2110 M2(4) = M2(1) + M2(2) + M2(3)
2120 C1 = M1(1) * M1(2) / M1(4)
2130 C2 = M1(1) * M1(3) / M1(4)
2140 C3 = M1(2) * M1(3) / M1(4)
2150 C4 = M2(1) * M2(2) / M2(4)
2160 C5 = M2(1) * M2(3) / M2(4)
2170 C6 = M2(2) * M2(3) / M2(4)
2180 C7 = C5 * I1 - C2 * I2
2190 C8 = (C1 + C2) * C5 - (C4 + C5) * C2
2200 C9 = (C2 + C3) * C5 - (C5 + C6) * C2
2210  RETURN
3000  REM  CALCULATE P=f(P1)*f(P2)
3010 T = 1:X = P1
3020 K1 = (C7 - C9 * X * X) / C8
3030 PX = (C1 + C2) * K1 + (C2 + C3) * X * X + 2 * C2 *  SQR (K1) * X - I
    1
3040  IF T = 2 GOTO 3070
3050 FA = PX:X = P2:T = 2
3060  GOTO 3020
3070 P = FA * PX
3080  RETURN
4000  REM  BISECTION ROUTINE
4010 A = 0.5:B = 5.0:FF = 0:IT = 0:H = B - A:E = 0.0001: REM  INITIAL PAR
    AMETERS
4020 P1 = A:P2 = B: GOSUB 3000: REM  P= f(a)*f(b)
4030  IF 0 >  = P THEN  GOTO 4070
4040  PRINT : PRINT "NO SOLUTION BETWEEN ";A;" AND ";B;"ANGSTROMS"
4050 FF = 1: REM  FAIL FLAG
4060  GOTO 4150: REM  EXIT ROUTINE
4070 I = I + 1:H = H / 2:C = A + H
4080 P1 = A:P2 = C: GOSUB 3000: REM  P=f(a)*f(c)
4090  IF P > 0 THEN  GOTO 4110
4100 B = C: GOTO 4120
4110 A = C
4120 E1 = H / 2
4130  IF E1 > E THEN  GOTO 4070
4140 R2 = (A + B) / 2:R1 =  SQR ((C7 - C9 * R2 * R2) / C8)
4150  RETURN
```

```
LINEAR TRIATOMIC DATA


 MOLECULE = OCS

FIRST ISOTOPE DATA
MASS 1 (AMU) = 16
MASS 2 (AMU) = 12
MASS 3 (AMU) = 32

ROTATIONAL CONSTANT (cm-1) = 0.202864

SECOND ISOTOPE DATA
MASS 1 (AMU) = 16
MASS 2 (AMU) = 12
MASS 3 (AMU) = 34

ROTATIONAL CONSTANT (cm-1) = 0.19791


**********

I1 = 8.30979219E-22 Kg m2 mol-1

I2 = 8.51779943E-22 Kg m2 mol-1

R12 = 1.171 ANGSTROMS

R23 = 1.552 ANGSTROMS
```

Program notes

(1) The program adopts a structured approach so that the controlling section of the program only occupies lines 100–170.
(2) The equation being solved is given in line 3030, although most of the constant terms are evaluated in lines 2100–2200.
(3) The bisection limits A,B are set in line 4010 along with E, the convergence accuracy required.
(4) The bisection routine sets a fail flag (FF = 1) if no solution has been found. This represents a way in which the main program can determine if the bisection routine has found a solution or not (line 140).

3.3 Polyatomic molecules

The rotational spectra of non-linear polyatomics are rather more complex and a detailed treatment is not feasible here. However, some general principles apply to all molecules and certain classes of polyatomic molecules exhibit fairly simple rotational spectra. In particular these molecules must possess symmetry properties which will enable us to simplify the description of their rotational motions.

3.3.1 *Classical rotation*

The classical energy of rotation of a rigid body is given by

$$E_{rot} = \tfrac{1}{2}\sum m_i(\dot{x}_i^2 + \dot{y}_i^2 + \dot{z}_i^2) \tag{3.31}$$

It can be shown, after considerable algebraic manipulation, that this can be represented as the rotation of the rigid body about three independent *principal axes* passing through the centre of mass of the body. Each principal axis has associated with it a moment of inertia and angular velocity.

$$E_{rot} = \tfrac{1}{2}I_a\omega_a^2 + \tfrac{1}{2}I_b\omega_b^2 + \tfrac{1}{2}I_c\omega_c^2 \tag{3.32}$$

It is conventional to order the moments of inertia such that $I_a < I_b < I_c$. Depending on the relative sizes of I_a, I_b, and I_c we can define certain classes of molecule, as shown in *Table 3.1*.

Table 3.1 The classification of rigid rotors based on their principal moments of inertia

Condition	*Type*	*Example*
$I_a = I_b = I_c$	Spherical top	CH_4, SF_6
$I_a = I_b < I_c$*	Oblate symmetric top	BCl_3, C_6H_6
$I_a < I_b = I_c$*	Prolate symmetric top	CH_3I, CF_3CCH
$I_a = 0$, $I_b = I_c$	Linear	HCl, OCS
$I_a \neq I_b \neq I_c$	Asymmetric top	H_2O, CH_2Br_2

* Some authors define I_a to be the unique axis for the symmetric top regardless of its size relative to I_b and I_c. Care must be taken to determine which convention is being used

The spherical top molecules are of no concern here since they have no permanent dipole moment and hence no pure rotational spectrum. The linear molecules have already been described in Section 3.2 and the asymmetric top molecules only possess closed, analytical expressions for their lowest energy levels and will not be dealt with here. This leaves us with the two *symmetric top* cases. These both possess one unique, and two equivalent, moments of inertia. The *oblate* symmetric top resembles a 'Frisbee' or flying saucer, whilst the *prolate* symmetric top has the shape of a rugby ball or cigar.

Fortunately many molecules can be described as near-prolate ($I_a < I_b \approx I_c$), or near-oblate ($I_a \approx I_b < I_c$), and a satisfactory explanation of their spectra can be obtained from the limiting case. For instance, formaldehyde, CH_2O, has three principal moments of inertia: $I_a = 2.98 \times 10^{-47}$ kg m^2, $I_b = 21.65 \times 10^{-47}$ kg m^2 and $I_c = 24.62 \times 10^{-47}$ kg m^2. Formaldehyde can thus be adequately described as a near-prolate symmetric top molecule.

The calculation of the three principal moments of inertia is difficult since we must first determine the locations of the three principal axes. There is a systematic (but complex) method for doing this for any molecule, but for symmetric top molecules the axes can usually be located easily by simple symmetry considerations. *Figure 3.6* shows the three principal axes for two symmetric top molecules, BF_3 and benzene.

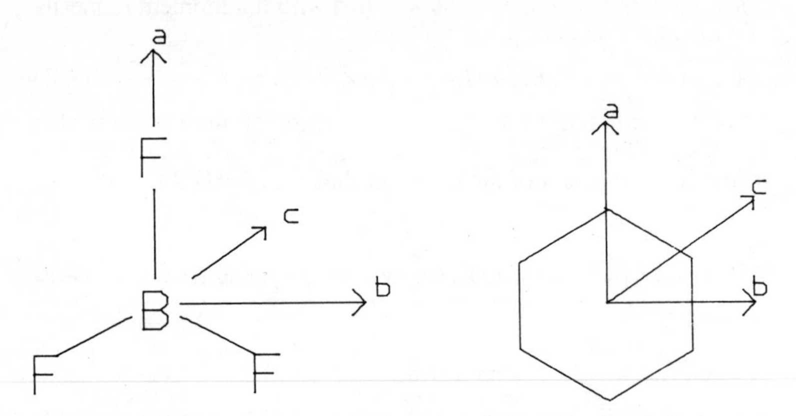

Figure 3.6 The principal axes for the symmetric top molecules BF_3 and benzene

3.3.2 *Symmetric tops—quantum solution*

The quantum solutions for the energy levels of symmetric top molecules are obtained by constructing the hamiltonian operator from Equation (3.32) and solving the Schrödinger equation. The energy levels for the prolate top ($I_b = I_c$) have the general form

$$E_{rot} = BJ(J+1) + (A-B)K^2 \tag{3.33}$$

and for the oblate top ($I_a = I_b$)

$$E_{rot} = BJ(J+1) + (C-B)K^2 \tag{3.34}$$

where $A = \dfrac{h^2}{8\pi^2 I_a}$, $B = \dfrac{h^2}{8\pi^2 I_b}$ and $C = \dfrac{h^2}{8\pi^2 I_c}$ joules.

As in Equation (3.10), dividing by $100hc$ will convert these to units of cm^{-1}.

We now find that there are two quantum numbers, J and K with values such that:

$$J=0, 1, 2\ldots \qquad K=0, \pm 1, \pm 2, \ldots, \pm J$$

The quantum number, J, determines the *total* angular momentum of the molecule while K gives the *component* of the angular momentum along the unique axis (I_a for prolate tops and I_c for oblate tops). We can thus see why K cannot exceed J. The positive and negative values of K arise since the direction of the angular momentum can be clockwise or anticlockwise about this axis.

An important difference between Equations (3.33) and (3.34) is evident if we consider the coefficient of K^2 in each case. For a prolate top molecule, $(A-B)$ is *positive*, and for a given J value the energy increases with increasing K. In the case of an oblate top $(C-B)$ is *negative* and, for a given J, the higher the K value the lower the energy.

In order for a symmetric top molecule to exhibit a rotational spectrum it must possess a permanent dipole moment. Most symmetric top molecules have this dipole moment directed along the unique axis—if this is the case then we find the following selection rules

$$\Delta J=0, \pm 1 \qquad \Delta K=0 \qquad (K\neq 0)$$

$$\Delta J=\pm 1 \qquad \Delta K=0 \qquad (K=0)$$

The fact that transitions between different K levels do not arise from photon absorption is not a surprise since the K quantum number signifies rotation about the unique axis—and hence the dipole. In this situation the rotation does not cause a change in dipole moment and hence it cannot interact with the electromagnetic field.

In an absorption experiment we have $\Delta J=+1, \Delta K=0$ and we find from Equations (3.33) or (3.34) that

$$\Delta E_J=2B(J+1) \tag{3.35}$$

which is exactly the same form as Equation (3.16). In other words, the line spacing for a symmetric top molecule looks just like those of a linear molecule. The degeneracies of the levels are somewhat different however. Each J level still has $(2J+1)$ possible orientations as before, corresponding to the different possible projections of J on a space fixed axis. However each J level also has $(2J+1)$ K values associated with possible projections on the (molecule fixed) unique axis. Thus, each (J, K) level is $(4J+2)$ degenerate for $K\neq 0$ and $(2J+1)$ if $K=0$. If we sum over all K levels for a given J level then each J level is $(2J+1)^2$ degenerate.

Program 3.4 calculates the J, K energy levels for a prolate top given the two principle moments of inertia.

Program 3.4 PROLE: Prolate top energy levels

```
100  REM  PROLATE TOP ENERGY LEVELS
110  GOSUB 1000: REM  GET DATA
120  PRINT : PRINT "ROTATIONAL CONSTANTS FOR: ";MN$
130  PRINT : PRINT "A = "; INT (AW * 1000 + .5) / 1000;" cm-1"
140  PRINT : PRINT "B = "; INT (BW * 1000 + .5) / 1000;" cm-1"
150  PRINT : PRINT "J     K             E/cm-1"
160  FOR J = 0 TO 3
170  PRINT
180  FOR K = 0 TO J
190 EW = BW * J * (J + 1) + (AW - BW) * K * K
200 EW =  INT (EW * 1000 + .5) / 1000
210  IF K <  > 0 THEN  GOTO 240
220  PRINT J;"     ";K,EW
230  GOTO 250
240  PRINT "      ";K,EW
250  NEXT : NEXT
260  END
1000  REM  GET MOMENTS OF INERTIA
1010  PRINT : PRINT "PROLATE TOP DATA": PRINT
1020  PRINT : INPUT "MOLECULE NAME = ";MN$
1030  PRINT : INPUT "IA (kg m2 mol-1) = ";IA
1040  PRINT : INPUT "IB (kg m2 mol-1) = ";IB
1050  IF IA < IB THEN  GOTO 1080
1060  PRINT : PRINT : PRINT "ERROR IA>IB - THIS IS AN OBLATE TOP": PRINT

1070  GOTO 1010: REM  TRY AGAIN
1080 H = 6.6256E - 34:L = 6.0225E23:C = 2.9979E8
1090 PI = 4 *  ATN (1)
1100 BW = H * L / IB:AW = H * L / IA
1110 BW = BW / (800 * PI * PI * C):AW = AW / (800 * PI * PI * C)
1120  RETURN
```

```
PROLATE TOP DATA


MOLECULE NAME = CH3Br

IA (kg m2 mol-1) = 3.318E-23

IB (kg m2 mol-1) = 5.44E-22

ROTATIONAL CONSTANTS FOR: CH3Br

A = 5.081 cm-1

B = .31 cm-1

J    K          E/cm-1

0    0          0

1    0          .62
     1          5.391

2    0          1.859
     1          6.63
     2          20.942
```

```
3  0      3.719
   1      8.489
   2      22.802
   3      46.655
```

Program note

(1) The moments of inertia are required in units of kg m^2 mol^{-1}.

3.4 Conclusions

The rigid rotor approximation in quantum mechanics is remarkably successful at describing the observed rotational spectra of diatomic and polyatomic molecules. A simple knowledge of masses and molecular geometry is all that is required to predict the spectrum. Conversely, the observed spectrum can be used to infer the molecular geometry with high precision. Careful studies do reveal discrepancies between theory and experiment and we must look to the non-rigid rotor for an answer. In fact, as well as centrifugal distortion, the vibrational motion of molecules plays an important role here, a point we shall pursue in the next chapter.

3.5 Reference

1. Mason, J.C., *BASIC Numerical Methods*, Chapter 4, Butterworths (1983).

3.6 Further reading

1. Barrow, G.M., *Introduction to Molecular Spectroscopy*, McGraw-Hill (1962).
2. Sugden, T.M. and Kenney, C.N., *Microwave Spectroscopy of Gases*, Van Nostrand (1965).
3. Townes, C.H. and Schawlow, A.L., *Microwave Spectroscopy*, McGraw-Hill (1955).
4. Flygare, W.H., 'Microwave Spectroscopy', in *Techniques of Chemistry* (Weissberger, A. and Rossiter, B.W., eds) Vol IIIA, 439, Wiley-Interscience (1972).

PROBLEMS

(3.1) Calculate the moments of inertia for $H^{35}Cl$, $H^{37}Cl$ and $D^{35}Cl$ given that they all share the equilibrium bond length of 1.275 Å.
(3.2) Calculate the values of the first three rotational transitions for each of the molecules in Problem (3.1).
(3.3) The $J=0$ to $J=1$ transition in $^{12}C^{16}O$ occurs at 3.842 cm^{-1}. Calculate the rotational constant $\bar{B}$ and hence find the internuclear distance.
(3.4) From the data in Problem (3.3), calculate the wavenumber of the $J=12$ to $J=13$ transition, and find the most intense spectral line

at 500 K (assuming that only the Boltzmann population is important).

(3.5) The rotational spectrum of $^{79}Br^{19}F$ has lines spaced 0.7143 cm^{-1} apart. Calculate the number of rotations per second that the molecule undergoes in $J=0$ and $J=5$. (Hint: find I and use Equations (3.7) and (3.8).)

(3.6) Verify Equation (3.19) by performing the necessary differentiation.

(3.7) By expanding the exponential term in Equation (3.23) and neglecting squared terms and higher, show that we can write Equation (3.26) as

$$I_{\text{abs}}^{nm} \propto (J+1)^3 \, e^{-BJ(J+1)/kT}$$

(3.8) Find an expression, analogous to Equation (3.19), which yields the J value with maximum intensity fot the expression in Problem (3.7). (Hint: remember only positive J values are physically meaningful.)

(3.9) Write a program that reads in the masses of a diatomic molecule and the rotational line spacing and outputs I, B, r, ω, and J_{max}.

(3.10) The following data apply to the rotational transitions of HCl

$4 \leftarrow 3$	83.03 cm^{-1}
$6 \leftarrow 5$	124.3
$8 \leftarrow 7$	165.51
$10 \leftarrow 9$	206.38

Calculate the values you would expect from Equation (3.16) with $\bar{B}=10.34$ cm^{-1} and compare them with those obtained from Equation (3.28) with $\bar{B}=10.395$, $\bar{D}=0.0004$.

(3.11) The $J=0$ to $J=1$ transition of $H^{19}F$ has a line spacing of 41.11 cm^{-1} whilst the $J=10$ to $J=11$ has a value of 37.81 cm^{-1}. Calculate the bond length of HF from the two transitions. Why are they different?

(3.12) Verify that for a linear triatomic molecule

$$I=\frac{m_1 m_2 r_{12}^2 + m_1 m_3 r_{13}^2 + m_2 m_3 r_{23}^2}{m_1+m_2+m_3} \qquad (3.22)$$

(Hint: the parallel axis theorem states that the moment of inertia I', about an axis parallel to the axis passing through the centre of gravity but displaced a distance R from it, is related to I by

$$I = I' + MR^2 \qquad (M=m_1+m_2+m_3)$$

We can thus calculate I' about one end atom and use the above relationship to find I, provided we also calculate the position of the centre of gravity, R, from the same end atom.)

(3.13) The lowest frequency microwave transitions of $H^{12}C^{14}N$ and $D^{12}C^{14}N$ lie at 88,631 MHz and 72,415 MHz respectively. Find the bond lengths in HCN.

(3.14) Classify the following molecules as spherical, symmetric or asymmetric tops:

$SiCl_4$, CH_3CN, XeF_4, NH_3, H_2S, HCN and *trans* SF_4Br_2.

(3.15) Modify Program 3.4 to input all three moments of inertia and hence calculate the energy levels of the *oblate* or *prolate* top molecule, depending on the moments of inertia.

(3.16) Plot the approximate (J, K) energy levels for the first three J levels of ethylene, given that its principal moments of inertia are $I_a = 5.75\times10^{-47}$, $I_b = 28.08\times10^{-47}$ and $I_c = 33.85\times10^{-47}$ kg m^2.

(3.17) BF_3 is shown in *Figure 3.6*. Show that the moments of inertia about the three principal axes are given by

$$I_a = I_b = \frac{3}{2} m_F r_{BF}^2, \qquad I_c = 3m_F r_{BF}^2$$

(3.18) Show that for an axial symmetric molecule ZXY_3 (such as CH_3Cl), the moments of inertia parallel and perpendicular to the symmetry axis are given by

$$I_{\parallel} = 2m_y R_{xy}^2(1-\cos\theta)$$

$$I_{\perp} = m_y R_{xy}^2(1-\cos\theta) + \left(\frac{m_y}{M}\right)(m_x+m_z)R_{xy}^2(1+2\cos\theta)$$

$$+\frac{m_z}{M} R_{xz}\{(3m_y+m_z)R_{xy} + 6m_y(\tfrac{1}{3}(1+2\cos\theta))^{\frac{1}{2}}\}$$

where $M = 3m_y + m_x + m_z$ and θ is the YXY bond angle.

(3.19) If we include centrifugal distortion in our discussion of the energy levels of symmetric tops then Equation (3.35) becomes

$$\Delta\bar{E}_{J,K} = 2\bar{B}(J+1) - 4\bar{D}_J(J+1)^3 - 2\bar{D}_{JK}(J+1)K^2$$

For CH_3F we find $\bar{B} = 0.851204$ cm^{-1}, $\bar{D}_J = 2.00\times10^{-6}$ cm^{-1} and $\bar{D}_{JK} = 1.47\times10^{-5}$ cm^{-1}. Compare the values obtained from this formula with the values you would get from Equation (3.33) and the following data: HCH bond angle $= 110°$, CH bond distance $= 1.109$ Å and CF bond distance $= 1.385$ Å. (Hint: for the second part you will have to calculate I_a and I_b first, see Problem (3.18).)

Chapter 4

Vibrational spectra

ESSENTIAL THEORY

The vibrational motion of molecules leads to absorption in the infra-red region of the spectrum. We again adopt the approach that proved so successful for molecular rotations; first we consider the classical description of the vibrational motion of a diatomic molecule, followed by the quantum mechanical solution to the problem. This will then be extended to polyatomic molecules and we shall also consider the case of simultaneous vibration and rotation, the so-called vib-rot problem.

In fact a comprehensive treatment of molecular vibrations is far too ambitious a task to be achieved in such a short book and we shall have to be content with a more restricted description than was possible with rotational motion.

4.1 Diatomic molecules

We can treat a vibrating diatomic molecule as two masses, m_1 and m_2, joined by a weightless spring. However, an even simpler mechanical case is that of a *single* mass, m, attached to a weightless spring, it is this latter case that we shall examine first.

4.1.1 *A single mass connected to a spring*

We assume that the force exerted on the mass by the spring is given by Hooke's law

$$f = -kq \tag{4.1}$$

where k is a *force constant* and q is the displacement of the mass from the equilibrium length of the spring. The force constant is essentially a measure of the stiffness of the spring and the minus sign arises because the force (a vector) is directed in the opposite direction to the displacement—i.e. it is a restoring force. Since an ideal spring is a *conservative* system (conserves energy) then we can obtain the potential energy, U, from

$$\frac{dU}{dq} = -f = kq \tag{4.2}$$

which yields on integration

$$U = \tfrac{1}{2}kq^2 \tag{4.3}$$

We can apply Newton's laws of motion very simply to the problem

$$f = ma = m\frac{d^2q}{dt^2} = -kq \tag{4.4}$$

The differential equation

$$\frac{d^2q}{dt^2} = \frac{-k}{m}q \tag{4.5}$$

is well known and has solutions of the form

$$q = q_0 \sin(2\pi\nu t + \phi) \tag{4.6}$$

where the vibrational frequency, ν, is given by

$$\nu = \frac{1}{2\pi}\sqrt{\frac{k}{m}} \tag{4.7}$$

The motion described by Equation (4.6) is called *simple harmonic motion.* The amplitude of the displacement from the equilibrium value ($q = 0$) varies sinusoidally with time and q_0 is the maximum displacement reached. The value of the phase shift ϕ depends on the starting conditions at $t = 0$.

4.1.2 *Two masses connected by a spring*

We now extend the treatment to include the case of two masses connected by a weightless spring. If the equilibrium length of the spring is r_e and the *instantaneous* length is given by r, then it is convenient to define a new variable

$$q = r - r_e \tag{4.8}$$

This represents the displacement of the spring from its equilibrium length. At present the treatment would include motion of the centre of mass of the molecule which makes the problem needlessly complicated since this motion has no effect on the vibrations themselves. We can remove this contribution simply be measuring the displacements of each mass relative to the centre of mass (*Figure 3.1*). This is achieved using Equation (3.4), following which, application of

Newton's laws to each mass produces

$$m_1 \frac{d^2 r_1}{dt^2} = -k(r - r_e)$$

and (4.9)

$$m_2 \frac{d^2 r_2}{dt^2} = -k(r - r_e)$$

substituting from Equations (3.4) and (4.8) we obtain just a single equation

$$\frac{m_1 m_2}{m_1 + m_2} \frac{d^2 r}{dt^2} = -kq \tag{4.10}$$

since r_e is a constant then $(d^2/dt^2)(r - r_e) = (d^2 r/dt^2)$ and we can write Equation (4.10) as

$$\frac{d^2 q}{dt^2} = \frac{-kq}{\mu} \tag{4.11}$$

where μ is the reduced mass (3.6). Equation (4.11) is identical to Equation (4.5) except that we have replaced m by μ. In other words, the two-particle problem is mathematically identical to the one-particle problem provided we use the reduced mass of the pair. This leads directly to the solution for the frequency of

$$\nu = \frac{1}{2\pi} \sqrt{\frac{k}{\mu}} \tag{4.12}$$

An important point to note is that only *one* vibrational frequency is allowed, although all amplitudes are possible.

4.1.3 *Quantum solution—the harmonic oscillator*

The hamiltonian operator for the harmonic oscillator problem requires that we be able to write down the classical total energy of the system, in this case two masses connected by a weightless spring.

$$H(q) = T(q) + U(q) \tag{4.13}$$

$$= \tfrac{1}{2} m_1 \dot{r}_1^2 + \tfrac{1}{2} m_2 \dot{r}_2^2 + \tfrac{1}{2} k q^2 \tag{4.14}$$

substituting for r_1 and r_2 from Equation (3.4) and remembering that $\dot{r}^2 = \dot{q}^2$ we obtain

$$H = \tfrac{1}{2} \mu \dot{q}^2 + \tfrac{1}{2} k q^2 \tag{4.15}$$

We can write this in terms of momentum $P = \mu \dot{q}$ for the 'single mass' system and apply the rules of *Table 2.1* to obtain the quantum mechanical hamiltonian operator

$$\hat{H} = -\frac{1}{8\pi^2\mu}\frac{d^2}{dq^2} + \tfrac{1}{2}kq^2 \quad (4.16)$$

The method of solving the resulting Schrödinger equation can be found in any standard textbook on quantum mechanics and will not be given here. All that we require are the solutions themselves.

4.1.4 *The energy levels*

It is again found that only certain, discrete values of energy are allowed and that these are determined by a single quantum number, V

$$E_V = (V + \tfrac{1}{2})\frac{h}{2\pi}\sqrt{\frac{k}{\mu}} \qquad V = 0, 1, 2\ldots \quad (4.17)$$

$$E_V = (V + \tfrac{1}{2})h\nu_{osc} \quad (4.18)$$

where $\nu_{osc} = (1/2\pi)\sqrt{(k/\mu)}$ is the frequency with which the system would vibrate classically. An important point to note from Equation (4.17) is that even for $V = 0$ the vibrational energy is non-zero. This residual *zero point energy* is a direct consequence of the Heisenberg uncertainty principle which places restrictions on the simultaneous observation of a particle's position and momentum.

Since cm^{-1} are still a convenient unit in the infra-red it is common to find the vibrational energy expressed in these units.

$$E_V = (V + \tfrac{1}{2})\left(\frac{\nu}{100c}\right) \qquad cm^{-1} \quad (4.19)$$

The quantity $\nu/100c$ is usually denoted by $\bar{\omega}_{osc}$ and is the vibrational frequency expressed in cm^{-1}. Note that there is a difference in terminology here from the rotational case where ω was used to denote an *angular* frequency, Equation (3.1).

4.1.5 *Wavefunctions*

It is instructive to look at the form of the wavefunctions themselves. The solutions involve the Hermite polynomials and have a general form:

$$\psi_V = N_V\, e^{-\frac{1}{2}\alpha q^2} H_V(\sqrt{\alpha}q) \quad (4.20)$$

where $\alpha^2 = (4\pi^2\mu k/h^2)$ and H_V is the Hermite polynomial of degree V.

The normalization constant, N_V, is given by

$$N_V = \left(\frac{1}{2^V V!} \sqrt{\frac{\alpha}{\pi}} \right)^{\frac{1}{2}} \tag{4.21}$$

The Hermite polynomials, for the first few values of V, are

$$\begin{aligned} H_0(y) &= 1 \\ H_1(y) &= 2y \\ H_2(y) &= 4y^2 - 2 \\ H_3(y) &= 8y^3 - 12y \\ H_4(y) &= 16y^4 - 48y^2 + 12 \end{aligned} \tag{4.22}$$

using Equations (4.20), (4.21) and (4.22) we obtain the first few

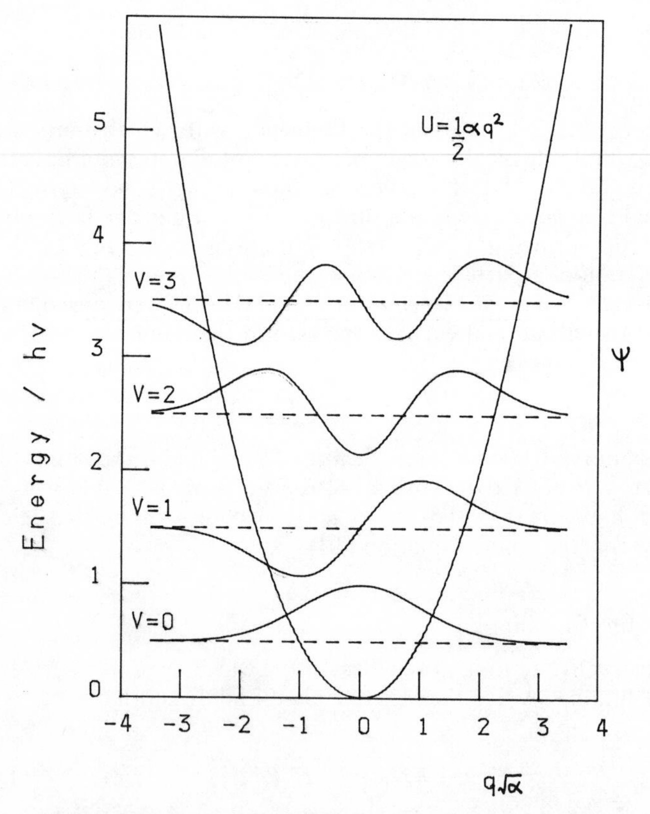

Figure 4.1 The harmonic oscillator energy levels and wavefunctions

solutions as

$$\psi_0 = \left(\frac{\alpha}{\pi}\right)^{\frac{1}{4}} . e^{-\frac{1}{2}\alpha q^2} \tag{4.23}$$

$$\psi_1 = \sqrt{2\alpha}\left(\frac{\alpha}{\pi}\right)^{\frac{1}{4}} q\, e^{-\frac{1}{2}\alpha q^2}$$

$$\psi_2 = \frac{1}{\sqrt{2}}\left(\frac{\alpha}{\pi}\right)^{\frac{1}{4}} (2\alpha q^2 - 1)\, e^{-\frac{1}{2}\alpha q^2}$$

Figure 4.1 shows the first four solutions superimposed on the harmonic potential energy curve. We can convert the horizontal scale to absolute lengths simply by dividing by $\sqrt{\alpha}$. The value of $\sqrt{\alpha}$ varies with bond strength and the diatomic masses, as defined in Equation (4.20). Values for HCl and CO are 9.1×10^{10} m^{-1} and 2.1×10^{11} m^{-1} respectively. Thus we find that $q\sqrt{\alpha} = 2$ represents a bond change of 0.22 Å for HCl but only 0.095 Å for CO.

Program 4.1 calculates the values of ψ and ψ^2 for the harmonic oscillator wavefunctions up to $V = 3$.

Program 4.1 PSIVIB: Diatomic vibrational wavefunctions

```
100  REM  VIBRATIONAL WAVEFUNCTIONS
110  GOSUB 1000: REM  GET DATA
120  PRINT : PRINT "CALCULATED VALUES FOR ";MN$
130  PRINT : PRINT "REDUCED MASS = ";MU;" kg"
140  PRINT : PRINT "FORCE CONSTANT = ";KF;" Nm-1"
150  PRINT : PRINT "WAVEFUNCTION FOR V = ";VL
160  REM  NOW DECIDE STEP SIZE
170 SS = 1 / (2 * RA): REM  FIRST APPROX
180  IF SS < 0.05E - 10 THEN SS = 0.05E - 10: GOTO 220
190  IF SS < 0.1E - 10 THEN SS = 0.1E - 10: GOTO 220
200  IF SS < 0.2E - 10 THEN SS = 0.2E - 10: GOTO 220
210 SS = 0.3: REM  LARGEST STEP SIZE
220  REM  NOW THE WAVEFUNCTION
230  PRINT : PRINT " q/A      Psi/m-1/2          Psi^2/m-1 ": PRINT
240  FOR I =  - 12 TO 12
250 Q = I * SS / 2: GOSUB 2000: REM  CALC PSI AND PSI^2
260 QA = Q * 1E10: REM  ANGSTROMS
270  PRINT QA; TAB( 9);P1; TAB( 25);P2
280  NEXT
290  END
1000  REM  GET DIATOMIC DATA
1010  PRINT : INPUT "MOLECULE = ";MN$
1020  PRINT : INPUT "MASS 1 = ";M1
1030  PRINT : INPUT "MASS 2 = ";M2
1040 MU = M1 * M2 / (M1 + M2)
1050  PRINT : INPUT "VIBRATIONAL FREQUENCY (cm-1) = ";W
1060 H = 6.6256E - 34:L = 6.0225E23:C = 2.9979E8
1070 PI = 4 *  ATN (1)
1080 MU = MU / (1000 * L): REM  kg
```

```
1090 VF = W * 2.998E10: REM  Hz
1100 KF = MU * ((2 * PI * VF) ^ 2)
1110 AL = 4 * PI * PI * MU * KF
1120 AL =  SQR (AL) / H: REM  m-2
1130 RA =  SQR (AL): REM  ROOT ALPHA
1140  PRINT : INPUT "VIB LEVEL (0-3) = ";VL
1150  IF VL < 0 OR VL > 3 THEN  GOTO 1140
1160 VM = VL + 1: REM   NUMBER FROM ONE
1170  RETURN
2000  REM  CALCULATE PSI AND PSI^2
2010 C1 =  SQR ( SQR (AL / PI)):C2 = RA * Q
2020 EQ =  EXP ( - (C2 * C2) / 2)
2030  ON VM GOTO 2040,2060,2080,2100
2040 P1 = C1 * EQ:P2 = P1 * P1
2050  GOTO 2110
2060 P1 = C1 * C2 *  SQR (2) * EQ:P2 = P1 * P1
2070  GOTO 2110
2080 P1 = (C1 * (2 * C2 * C2 - 1) * EQ) /  SQR (2)
2090  GOTO 2110
2100 P1 = (C1 * (2 * C2 * C2 * C2 - 3 * C2) * EQ) /  SQR (3)
2110 P2 = P1 * P1
2120  RETURN


MOLECULE = CO

MASS 1 = 12

MASS 2 = 16

VIBRATIONAL FREQUENCY (cm-1) = 2143

VIB LEVEL (0-3) = 0

CALCULATED VALUES FOR CO

REDUCED MASS = 1.13858744E-26 kg

FORCE CONSTANT = 1855.38365 Nm-1

WAVEFUNCTION FOR V = 0

 q/A      Psi/m-1/2          Psi^2/m-1

-.3       1.0405873E-03   1.08282193E-06
-.275     .0238690959     5.69733738E-04
-.25      .416950087      .173847375
-.225     5.54654858      30.7642012
-.2       56.1891209      3157.2173
-.175     433.48324       187907.719
-.15      2546.73079      6485837.69
-.125     11394.2154      129828145
-.1       38821.8715      1.5071377E+09
-.075     100730.077      1.01465485E+10
-.05      199036.408      3.96154916E+10
-.025     299499.802      8.97001313E+10
0         343203.107      1.17788373E+11
.025      299499.802      8.97001313E+10
.05       199036.408      3.96154916E+10
.075      100730.077      1.01465485E+10
```

```
.1      38821.8715      1.5071377E+09
.125    11394.2154      129828145
.15     2546.73079      6485837.69
.175    433.48324       187907.719
.2      56.1891209      3157.2173
.225    5.54654858      30.7642012
.25     .416950087      .173847375
.275    .0238690959     5.69733738E-04
.3      1.0405873E-03   1.08282193E-06
```

Program notes

(1) Masses are input in amu and frequencies in cm^{-1}. These are converted to S.I. quantities in lines 1080–1130.
(2) An appropriate step size is calculated in lines 170–210 utilizing the fact that $\sqrt{\alpha}q = \pm 3$ will cover most of the wavefunction.
(3) Lines 240–280 are responsible for the main calculation.
(4) The first four solutions ($V=0-3$) are stored explicitly in the routine at 2000. The required solution is accessed using the ON GOTO statement.
(5) The shape of ψ can usually be seen by observing the position of the decimal point in the output.

4.1.6 *Selection rules and line intensities*

Selection rules for transitions are found by evaluating the transition dipole moment integral (2.32). Substitution of the harmonic oscillator wavefunctions (4.20) yields the result

$$\Delta V = \pm 1 \tag{4.24}$$

We can partly understand how this arises by looking at the symmetry of the functions concerned. The transition moment integral has the explicit form

$$R_{nm} = e \int_{-\infty}^{\infty} \psi_n^* q \psi_m \, dq \tag{4.25}$$

An examination of the wavefunctions (*Figure 4.1*) shows that $V=0$, 2, 4 ... are all *even* functions of q, while $V=1$, 3, 5 ... are all *odd*. To ensure that Equation (4.24) has a non-zero value, the product $\psi_n^* q \psi_m$ must be even overall. Since q itself is an odd function this requires that ψ_n and ψ_m be made up of one odd and one even function. We would thus expect purely on symmetry grounds a rule of the form $\Delta V = \pm 1$, ± 3.... The exact treatment eliminates all but the first possibility.

Given the rule in Equation (4.24) and the energy level formula (4.18) it is easy to show that *every* transition in a vibrational spectrum has the same energy (since the levels are equidistant apart). The simple harmonic oscillator model thus predicts a vibrational spectrum consisting of a *single* line made up of the superposition of all the allowed transitions. In fact, as we shall see later, this is not quite the case since real molecules do not vibrate in a truly harmonic potential. This has the effect of producing a closely spaced set of vibrational lines rather than just a single one.

The intensities of these lines can be calculated from the Boltzmann populations of the individual levels. In fact, in contrast to rotation, this is the only factor which we need to consider in Equation (2.34) or (2.36) to obtain relative intensities. We find that at room temperature most molecules only have significant populations for $V=0$ and $V=1$—very different from the rotational case. This results from two factors: first, the energy level spacing for vibration is much larger than for rotation, and secondly the vibrational levels are non-degenerate.

For instance, the vibrational frequency of carbon monoxide is 2170 cm^{-1} (4.31×10^{-20} J), which at 300 K produces

$$\frac{n_1}{n_0}=e^{-(E_1-E_0)kT}=3\times10^{-5}$$

In this case essentially all the molecules are in the ground vibrational level. If, however, we consider the much less rigid molecule I_2 with a vibrational spacing of only 214 cm^{-1} we obtain a value for n_1/n_0 of 0.358. Here there are significant (but still minor) populations for $V=1$ and $V=2$.

4.1.7 *Force constants*

The force constant, k, is a measure of the 'stiffness' of the spring or bond between the atoms. We can obtain a value for the force constant from the observed energy level spacing via Equation (4.17).

$$k=\left(\frac{2\pi\,\Delta E}{h}\right)^2 \qquad (4.26)$$

For CO, ΔE, the energy gap between successive levels, is 4.31×10^{-20} J, and $\mu=1.138\times10^{-26}$ kg. This produces a value for k of 1902 Nm^{-1}. Carbon monoxide is a very stiff bond and this is reflected in the large value for the force constant. On the other hand, the less rigid molecule I_2 with an energy spacing of 4.25×10^{-21} J has a force constant of only 171 Nm^{-1}. We can see that as a bond becomes less stiff the energy level spacing decreases and the thermal population of the higher V levels increases. *Table 4.1* lists vibrational frequencies and the resulting force constants for several common diatomic molecules.

Table 4.1 Diatomic vibrational frequencies and force constants

Molecule	$\bar{\omega}$/cm^{-1}	k/Nm^{-1}
H_2	4159	509
HF	3958	877
HCl	2890	478
HBr	2559	381
Cl_2	557	320
N_2	2331	2241
NaCl	378	117

The inclusion of homonuclear diatomic molecules in the list seems to contradict the requirement of a permanent dipole moment, however those results have been obtained by vibrational *Raman* spectroscopy.

4.2 Anharmonicity

4.2.1 *Improved molecular potentials*

The model of two point masses joined by a spring obeying Hooke's law is obviously an oversimplification of the situation in real molecules. True molecules dissociate if we stretch a bond enough and, going in the other direction, exhibit enormous repulsive energy as we try to push the nuclei into each other. In order to improve the description of the vibrational motion of diatomic molecules we must change the potential energy function to reflect this asymmetry.

Figure 4.2 shows a typical potential energy curve for a diatomic molecule. It also includes several different approximations to this curve and we now examine where they arise and how successful they are at describing real vibrational spectra.

As a starting point, one can expand the true potential energy function $U(q)$ in a Maclaurin series about $q=0$.

$$U(q)=U_{q=0}+\left(\frac{\mathrm{d}U}{\mathrm{d}q}\right)_{q=0}q+\frac{1}{2}\left(\frac{\mathrm{d}^2U}{\mathrm{d}q^2}\right)_{q=0}q^2+\frac{1}{3!}\left(\frac{\mathrm{d}^3U}{\mathrm{d}q^3}\right)_{q=0}q^3+\cdots \tag{4.27}$$

Since we are free to choose a zero of energy it simplifies matters if we choose it to be the bottom of the potential well, $U_{q=0}=0$. Further, since the bottom of the well is a minimum, we know $(\mathrm{d}U/\mathrm{d}q)_{q=0}=0$

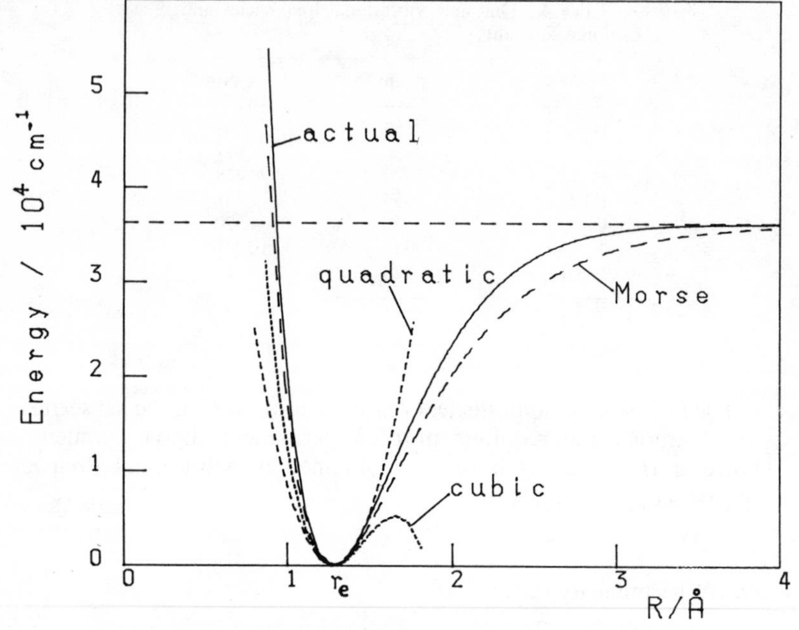

Figure 4.2 The potential energy curve for a diatomic molecule and several approximations to it

and Equation (4.27) becomes

$$U(q)=\frac{1}{2}\left(\frac{\mathrm{d}^2U}{\mathrm{d}q^2}\right)_{q=0}q^2+\frac{1}{3!}\left(\frac{\mathrm{d}^3U}{\mathrm{d}q^2}\right)_{q=0}q^3+\cdots \qquad (4.28)$$

Truncating this expansion at q^2 produces the familiar harmonic oscillator potential where $k=(\mathrm{d}^2U/\mathrm{d}q^2)_{a=0}$.

Keeping terms up to q^3 provides an improved fit close to the bottom of the well but a cubic equation has two turning points so the curve must turn over again in a physically unrealistic manner at larger distances. It is possible to keep adding higher order terms to the expansion to obtain a better fit to the potential but the difficulty in solving the resulting Schrödinger equation increases dramatically. In fact, even if we stop at the cubic stage we must use perturbation techniques to obtain the energy levels.

4.2.2 *Anharmonic energy levels*

When the Schrödinger equation is solved using a cubic potential one obtains energy levels of the form

$$\bar{E}_V = \bar{\omega}_e(V+\tfrac{1}{2}) - \overline{\omega_e x_e}(V+\tfrac{1}{2})^2 \text{ cm}^{-1} \tag{4.29}$$

The $\overline{\omega_e x_e}$ term, called the *anharmonicity constant*, is positive and much smaller than $\bar{\omega}_e$. Although it is a product of $\bar{\omega}_e$ and $\bar{x}_e$ it is unusual to see the $\bar{x}_e$ value given separately and throughout this book we shall use $\overline{\omega_e x_e}$ as a single number. Because the anharmonicity constant is positive the anharmonic oscillator levels exhibit decreasing separation with increasing V and *converge* as they approach the dissociation limit. However, unlike the hydrogen atom there is a finite number of such levels leading to dissociation.

Comparing Equation (4.29) with (4.19) one is tempted to equate $\bar{\omega}_e$ with an $\bar{\omega}_{osc}^{anh}$ since they are both coefficients of the $(V+\tfrac{1}{2})$ term, however, although $\bar{\omega}_{osc}^{anh}$ and $\bar{\omega}_e$ are related they are not identical. We can rewrite Equation (4.29) as

$$\bar{E}_V = \bar{\omega}_e[1 - \bar{x}_e(V+\tfrac{1}{2})] \, . \, (V+\tfrac{1}{2}) \tag{4.30}$$

comparison with Equation (4.19) suggests we could write

$$\bar{\omega}_{osc}^{anh} = \bar{\omega}_e[1 - \bar{x}_e(V+\tfrac{1}{2})] \tag{4.31}$$

If we now consider the hypothetical energy state $V = -\tfrac{1}{2}$ $(\bar{E}_V = 0)$ its oscillation frequency would be given by $\bar{\omega}_{osc}^{anh} = \bar{\omega}_e$. Thus $\bar{\omega}_e$ is the (hypothetical) equilibrium oscillation frequency—the frequency of oscillation for infinitesimal motions about the bottom of the well. Using Equation (4.29) are (4.30) we find that the zero point energy of the anharmonic oscillator $(V=0)$ is given by

$$E_0 = \tfrac{1}{2}\bar{\omega}_e \left(1 - \frac{\bar{x}_e}{2}\right) \qquad \text{cm}^{-1} \tag{4.32}$$

The selection rule for the anharmonic oscillator turns out to be much less restricting than for the simple harmonic oscillator and it now becomes

$$\Delta V = \pm 1, \pm 2, \pm 3 \ldots \tag{4.33}$$

Although all changes in V are allowed, it is generally found that the intensities of the spectral lines fall off rapidly above $\Delta V = \pm 1$, reflecting the partially 'forbidden' character of the higher transitions.

The experimentally observed spectral lines can be used to yield values for the constants $\overline{\omega}_e$ and $\overline{\omega_e x_e}$. Since lines corresponding to $\Delta V = \pm 1$ have similar values and thus lie rather close together it is often more convenient to use the *overtone* or *harmonic* lines ($\Delta V = \pm 2, \pm 3 \ldots$) to achieve this. Further, since most IR spectra are taken in absorption and at room temperature, only $V=0$ is significantly populated and it is the $V \leftarrow 0$ ($V=1, 2, 3 \ldots$) transitions that are used. The energy spacing is obtained from Equation (4.29).

$$\bar{E}_V - \bar{E}_0 = \bar{\omega}_e V - \overline{\omega_e x_e}\, V(V+1) \tag{4.34}$$

Program 4.2 uses a least squares technique[1] to provide the best estimate of $\overline{\omega}_e$ and $\overline{\omega_e x_e}$ from the $V \leftarrow 0$ overtone lines using Equation (4.34). *Table 4.2* provides such overtone data for HCl and the CH bond in $CHCl_3$.

Table 4.2 Fundamental and overtone vibrational frequencies for HCl and the CH bond in $CHCl_3$

ΔV	HCl ($\bar{\omega}$/cm^{-1})	CH in $CHCl_3$ ($\bar{\omega}$/cm^{-1})
$1 \leftarrow 0$	2886	3019
$2 \leftarrow 0$	5668	5900
$3 \leftarrow 0$	8347	8700
$4 \leftarrow 0$	10923	11315
$5 \leftarrow 0$	13397	13860

Program 4.2 OVDAT: Vibrational constants from overtone data

```
90  DIM X(10),Y(10)
100  REM  OVERTONE DATA PROGRAM
110  GOSUB 1000: REM  GET DATA
120  GOSUB 2000: REM   FIT TO Y=AX+BX^2
130 WX = B: REM  WeXe
140 WE = A + B: REM  We
150 WX =  INT (WX * 100 + .5) / 100
160 WE =  INT (WE * 100 + .5) / 100
170  PRINT
180  PRINT "We    = ";WE;" cm-1"
190  PRINT "WeXe = ";WX;" cm-1"
200  REM  NOW THE BACK FIT VALUES
210  PRINT : PRINT "EXP."; TAB( 12);"CALC."; TAB( 20);"DIFF."
220  FOR V = 1 TO N
230 EF = WE * V - WX * V * (V + 1)
240 EF =  INT (EF * 100 + .5) / 100
250 DF = Y(V) - EF
260 DF =  INT (DF * 100 + .5) / 100
270  PRINT Y(V); TAB( 10);EF; TAB( 20);DF
280  NEXT
290  END
1000  REM  OVERTONE DATA INPUT
1010  PRINT : PRINT "INPUT FUNDAMENTAL AND OVERTONE DATA"
1020  PRINT "(FREQUENCIES IN cm-1)"
1030  PRINT : INPUT "NUMBER OF LINES = ";N
1040  IF N < 2 THEN  PRINT "ERROR - MUST BE AT LEAST TWO": GOTO 1030
1050  PRINT
1060  FOR V = 1 TO N
```

```
1070  PRINT V;"<-0";: INPUT " FREQ = ";F
1080 Y(V) = F:X(V) = V
1090  NEXT
1100  RETURN
2000  REM   LEAST SQUARE FIT Y = AX + BX^2
2010  REM  N=NUMBER OF POINTS
2020  REM  X() AND Y() HOLD DATA
2030 S1 = 0:S2 = 0:S3 = 0:S4 = 0:S5 = 0
2040  FOR I = 1 TO N
2050 X1 = X(I):X2 = X1 * X1:X3 = X2 * X1:X4 = X2 * X2
2060 S1 = S1 + X2:S2 = S2 + X3:S3 = S3 + X4
2070 S4 = S4 + Y(I) * X1:S5 = S5 + Y(I) * X2
2080  NEXT
2090 DE = S2 * S2 - S1 * S3
2100 A = (S2 * S5 - S4 * S3) / DE
2110 B = (S1 * S5 - S4 * S2) / DE
2120  RETURN


INPUT FUNDAMENTAL AND OVERTONE DATA
(FREQUENCIES IN cm-1)

NUMBER OF LINES = 5

1<-0 FREQ = 2886
2<-0 FREQ = 5668
3<-0 FREQ = 8347
4<-0 FREQ = 10923
5<-0 FREQ = 13397

We   = 2988.52 cm-1
WeXe = 51.53 cm-1

EXP.        CALC.     DIFF.
2886        2885.46   .54
5668        5667.86   .14
8347        8347.2    -.2
10923       10923.48  -.48
13397       13396.7   .3
```

Program notes

(1) The dimensions in line 105 are sufficient to handle up to the $10 \leftarrow 0$ overtone transition, these can be increased if necessary.
(2) The routine at line 2000 performs a least squares fit to the equation $Y = aX + bX^2$ using the data in $X(\;)$, and $Y(\;)$.
(3) The experimental, calculated and difference values are output in lines 200–280.

The values obtained for $\overline{\omega_e}$ and $\overline{\omega_e x_e}$ are extremely useful since they summarize the anharmonicity in a very concise form. For some molecules two parameters are not sufficient to give a good description of the observed energy spacings and we must include a term in $(V + \frac{1}{2})^3$ in Equation (4.29). The coefficient of this term is denoted by

Table 4.3 Vibrational constants for the ground states of several diatomic molecules

Molecule	$\overline{\omega_e}$ /cm^{-1}	$\overline{\omega_e x_e}$ /cm^{-1}	$\overline{\omega_e y_e}$ /cm^{-1}
H_2	4401.2	121.3	0.812
D_2	3115.5	61.82	0.562
HCl	2990.9	52.82	0.2243
HBr	2648.9	45.22	−0.0029
HI	2309.0	39.64	−0.0200
Cl_2	559.7	2.67	−0.006
Br_2	325.3	1.0774	−0.0023
I_2	214.50	0.614	−0.000895
N_2	2358.7	14.324	−0.00226
O_2	1580.19	11.98	0.0474
CO	2169.81	13.288	0.0105

$\overline{\omega_e y_e}$. *Table 4.3* provides values for $\overline{\omega}_e$, $\overline{\omega_e x_e}$ and $\overline{\omega_e y_e}$ for several diatomic molecules.

Program 4.3 uses the $\overline{\omega}_e$ and $\overline{\omega_e x_e}$ values from *Table 4.3* to calculate the harmonic and anharmonic energy levels for a diatomic molecule. The spacings between the anharmonic energy levels, corresponding to the expected vibrational transitions, are also given. The harmonic spacing is assumed to be constant and equal to $\bar{E}_1 - \bar{E}_0$. The program also outputs the expected values of the overtone transitions in both the harmonic and anharmonic cases.

Program 4.3 VIBLEV: Diatomic vibrational levels

```
100  REM  VIBRATIONAL SPECTRUM
110  GOSUB 1000: REM  GET DATA
120  REM  CALCULATE HARMONIC We
130 WH = WE - 2 * WX
140  GOSUB 2000: REM  ENERGY LEVELS
150  GOSUB 3000: REM  OVERTONE TRANSITIONS
160  END
1000  REM  VIBRATIONAL CONSTANTS
1010  PRINT : PRINT "DIATOMIC CONSTANTS"
1020  PRINT "(ALL QUANTITIES IN cm-1)"
1030  PRINT : INPUT "MOLECULE = ";MN$
1040  PRINT : INPUT "We = ";WE
1050  PRINT : INPUT "WeXe = ";WX
1060  PRINT : INPUT "NUMBER OF LEVELS = ";N
1070  RETURN
2000  REM  ENERGY LEVELS
2010  PRINT : PRINT "V"; TAB( 5);"HARMONIC"; TAB( 15);"ANHARMONIC"; TAB(
   26);"ANH.SPACING"
2020  FOR V = 0 TO N
2030 EH = WH * (V + .5)
2040 EA = WE * (V + .5) - WX * ((V + .5) ^ 2)
2050  IF V = 0 THEN EL = EA: GOTO 2090
```

```
2060 DE = EA - EL
2070  PRINT  TAB( 27);DE: REM   ENERGY SPACING
2080 EL = EA: REM   UPDATE LOWER LEVEL
2090  PRINT V; TAB( 6);EH; TAB( 16);EA
2100  NEXT
2110  PRINT : PRINT "HARMONIC SPACING = ";WH
2120  RETURN
3000  REM  CALCULATE OVERTONE VALUES
3010  PRINT : PRINT : PRINT "OVERTONE TRANSITIONS"
3020  PRINT : PRINT "TRANS."; TAB( 9);"HARMONIC"; TAB( 19);"ANHARMONIC"
3030  FOR V = 1 TO N
3040 EH = V * WH
3050 EA = WE * V - WX * V * (V + 1)
3060  PRINT V;"<-0"; TAB( 10);EH; TAB( 20);EA
3070  NEXT
3080  RETURN
```

```
DIATOMIC CONSTANTS
(ALL QUANTITIES IN cm-1)

MOLECULE = HI

We = 2309

WeXe = 39.64

NUMBER OF LEVELS = 5

V   HARMONIC  ANHARMONIC ANH.SPACING
0    1114.86   1144.59
                          2229.72
1    3344.58   3374.31
                          2150.44
2    5574.3    5524.75
                          2071.16
3    7804.02   7595.91
                          1991.88
4    10033.74  9587.79
                          1912.6
5    12263.46  11500.39

HARMONIC SPACING = 2229.72


OVERTONE TRANSITIONS

TRANS.  HARMONIC  ANHARMONIC
1<-0     2229.72   2229.72
2<-0     4459.44   4380.16
3<-0     6689.16   6451.32
4<-0     8918.88   8443.2
5<-0     11148.6   10355.8
```

4.2.3 *The Morse potential*

In 1929, P.M. Morse[2] suggested an empirical form for the potential energy function $U(q)$. The function, which is shown in *Figure 4.2*, has two advantages over the power series approach to $U(q)$. First, it has the correct overall shape across the whole range of internuclear distances, and secondly it produces analytic solutions for the vibrational Schrödinger equation. The Morse curve has the form:

$$U(q) = D_e(1 - e^{-\beta q})^2 \qquad (4.35)$$

In S.I. units D_e is in joules, β in m^{-1} and q in m. Substitution of Equation (4.35) into the vibrational hamiltonian (4.13) produces energy levels given by

$$E_V = hc\omega_e(V + \tfrac{1}{2}) - hc\omega_e x_e(V + \tfrac{1}{2})^2 \quad \text{J} \qquad (4.36)$$

with no higher terms in $(V + \frac{1}{2})$ and where ω_e and $\omega_e x_e$ are in m^{-1}. Furthermore, β and ω_e are related by

$$\beta = \pi c \sqrt{\frac{2\mu}{D_e}} . \omega_e \quad m^{-1} \qquad (4.37)$$

where ω_e is the frequency expressed in m^{-1} and μ is the reduced mass in kg mol^{-1}. The anharmonicity constant is given by

$$\omega_e x_e = \frac{h\beta^2}{8\pi^2 c} \quad m^{-1} \qquad (4.38)$$

It is common to use the empirical value of D_e and ω_e to calculate a value for β from Equation (4.37). In principle the value of $\omega_e x_e$ can then be found from Equation (4.38). However, if the true potential curve is not well described by the Morse potential, the calculated anharmonicity will not agree with its empirically determined value.

Since S.I. quantities are still used infrequently in this spectral range we also provide Equations (4.37) and (4.38) in more common units

$$\beta = \sqrt{\frac{2\pi^2 c\mu_g}{10^5 h\bar{D}_e}} \bar{\omega}_e \quad cm^{-1}$$

and

$$\overline{\omega_e x_e} = \frac{10^5 h\bar{\beta}^2}{8\pi^2 c\mu_g} \quad cm^{-1} \qquad (4.39)$$

Here $\bar{\beta}$, $\bar{D}_e$ and $\bar{\omega}_e$ are all in cm^{-1} and the reduced mass, μ_g is in g mol^{-1}.

Although the Morse potential has the correct overall shape it still only produces an energy expression equivalent to the cubic expansion in Equation (4.28). As a result of this, the Morse potential is often not accurate enough for detailed spectroscopy and we must look for a better representation of the potential curve. Several other empirical potentials have been suggested which offer greater accuracy over a larger range but they are more complicated to deal with and they will not be discussed here. The books by Herzberg listed at the end of this chapter should be consulted for more detail.

4.2.4 *Dissociation limit and dissociation energy*

Since the energy levels for an anharmonic potential converge as they approach the dissociation limit there exists a maximum vibrational level that is accessible to the molecule. Furthermore, as can be seen in *Figure 4.3*, the dissociation energy D_0 can be found by summing all the vibrational spacings up to this final level.

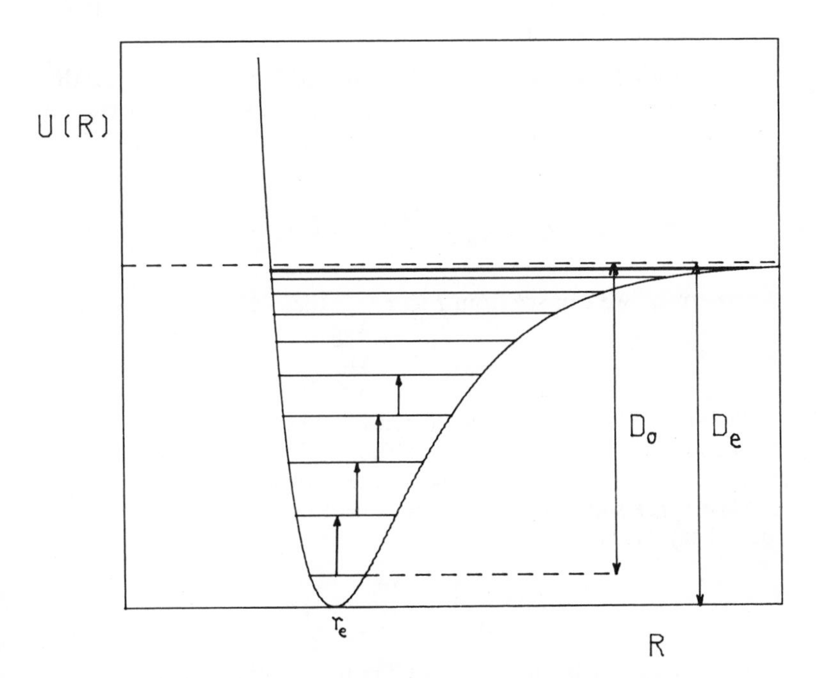

Figure 4.3 The $\Delta V = +1$ transitions in an anharmonic oscillator

For a Morse potential (or a cubic potential) we can write the energy levels as

$$\bar{E}_V = \bar{\omega}_e(V+\tfrac{1}{2}) - \overline{\omega_e x_e}\,(V+\tfrac{1}{2})^2 \qquad \mathrm{cm}^{-1} \qquad (4.40)$$

the *energy spacing* between levels then becomes

$$\Delta\bar{E}_V=\bar{E}_{V+1}-\bar{E}_V=\bar{\omega}_e(V+\tfrac{3}{2})-\overline{\omega_e x_e}\,(V+\tfrac{3}{2})^2$$
$$-\bar{\omega}_e(V+\tfrac{1}{2})+\overline{\omega_e x_e}\,(V+\tfrac{1}{2})^2.$$
$$\Delta\bar{E}_V=\bar{\omega}_e-2\overline{\omega_e x_e}\,(V+1) \qquad (4.41)$$

where V refers to the *lower* level of the pair.

We can find the V level corresponding to the dissociation limit by treating Equation (4.40) as a continuous function of V, at the dissociation limit the energy spacing goes to zero so $(\mathrm{d}E_V/\mathrm{d}V)=0$.

Differentiating Equation (4.40) and equating the result to zero produces the result

$$V_{\max}=\frac{\bar{\omega}_e}{2\overline{\omega_e x_e}}-\frac{1}{2} \qquad (4.42)$$

In reality we should truncate the result of Equation (4.42) to the nearest lower integer. From *Figure 4.3* it is easy to see that the energy of this level is just $\bar{D}_e$, and substituting Equation (4.42) into (4.40) we obtain

$$\bar{D}_e=\bar{E}_{V_{\max}}=\frac{\bar{\omega}_e^2}{4\overline{\omega_e x_e}} \qquad \mathrm{cm}^{-1} \qquad (4.43)$$

Alternatively we can see from *Figure 4.3* that

$$\bar{D}_0=\sum_{V=0}^{V_{\max}-1}\Delta\bar{E}_V \qquad (4.44)$$

The summation index must be considered carefully; it starts from zero and extends up to $V_{\max}-1$. This is so because the V values in Equation (4.41) represent the *lower* state of the transition, but $V_{\max}$ is the last *upper* state we could access. Substitution of Equation (4.41) into (4.44) yields

$$\bar{D}_0=\frac{\bar{\omega}_e}{4\overline{\omega_e x_e}}-\frac{\bar{\omega}_e}{2}+\frac{\overline{\omega_e x_e}}{4} \qquad (4.45)$$

We can obtain $\bar{D}_e$ simply by adding the zero point energy (4.32). This yields, as expected

$$\bar{D}_e=\frac{\bar{\omega}_e^2}{4\overline{\omega_e x_e}} \qquad \mathrm{cm}^{-1}$$

Finally, this result can also be verified by substituting expressions from the Morse parameters (4.39).

4.2.5 *The Birge–Sponer extrapolation*

The above formulae apply only if the energy levels are correctly described by Equation (4.40). However, it is common, especially for high V, to find that cubic and quartic terms must be included in Equation (4.40) and the dissociation energy is no longer given by Equation (4.43). In this case we must return to Equation (4.44) and obtain $\bar{D}_0$ by summing up all the individual transition energies. This approach will only be successful if the vibrational spectrum contains enough lines to observe the convergence limit directly. Unfortunately, especially in absorption spectra, one can only observe the first few vibrational lines.

An approximate method to overcome this difficulty was suggested by Birge and Sponer. In this we plot the $\Delta\bar{E}_V$ values versus $(V+\frac{1}{2})$ (*i.e.* $1 \leftarrow 0$ is plotted midway between $V=0$ and $V=1$) as shown in *Figure 4.4*

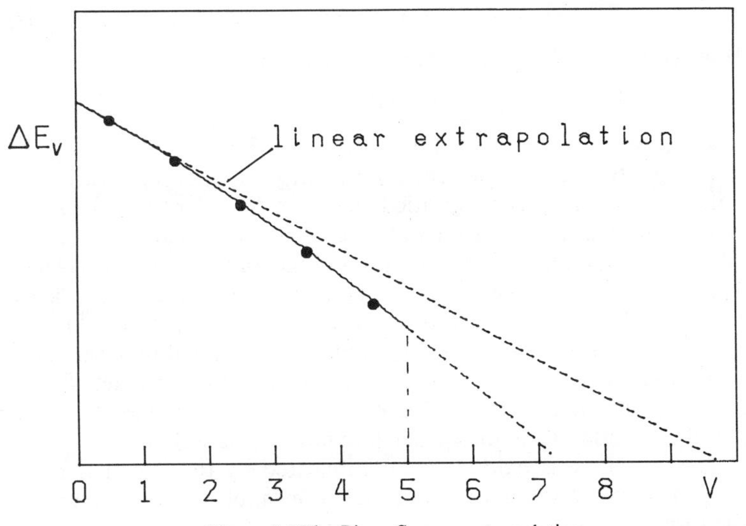

Figure 4.4 The Birge–Sponer extrapolation

We then extrapolate the curve until it meets the V axis ($\Delta\bar{E}_V=0$). The area under the curve is then very close to Equation (4.44) and it represents a significant improvement over the linearly extrapolated value corresponding to Equation (4.43)

$$D_0 \simeq \text{area under curve} \tag{4.46}$$

The majority of molecules exhibit negative curvature as shown in

Figure 4.4 but some molecules, mainly ionic species, have positive curvature. In the infra-red a combination of rapidly diminishing Boltzmann populations and a dominant $\Delta V = \pm 1$ selection rule means that generally only a few vibrational lines can be observed. As a result of this the Birge–Sponer extrapolation is more widely used with *electronic* spectra—which usually exhibit very many more vibrational lines.

At first glance it seems a relatively simple task to write a program that performs the Birge–Sponer extrapolation and integration automatically. In fact, as a little thought soon shows, it turns out to be a somewhat difficult problem with several facets. This provides an opportunity to explore a numerical problem more fully and the method adopted here is by no means the only one.

In order to introduce curvature into the Birge–Sponer plot we must retain terms at least up to $(V+\frac{1}{2})^3$ in the energy level expression. We shall assume here that the cubic term will be sufficient for our purposes

$$\bar{E}_V = \overline{\omega}_e(V+\tfrac{1}{2}) - \overline{\omega_e x_e}\,(V+\tfrac{1}{2})^2 + \overline{\omega_e y_e}\,(V+\tfrac{1}{2})^3 \qquad (4.47)$$

If we consider $\Delta V = +1$ transitions this yields

$$\Delta\bar{E}_V = a + b(V+\tfrac{1}{2}) + c(V+\tfrac{1}{2})^2 \qquad V = 0, 1, 2 \ldots \qquad (4.48)$$

where $a = \overline{\omega}_e - \overline{\omega_e x_e} + \overline{\omega_e y_e}$, $b = -(2\overline{\omega_e x_e} - 3\overline{\omega_e y_e})$ and $c = 3\overline{\omega_e y_e}$.

The first two terms represent a linear Birge–Sponer plot, whilst the third term contributes quadratic behaviour. In order to perform the extrapolation and integration numerically we have to fit the observed vibrational transitions to an equation of the form of Equation (4.48). This involves using the observed data to determine the best (least squares) fit for the coefficients *a*, *b*, *c*. Performing a least squares fit with three coefficients involves considerable mathematical manipulation and the reader is referred to References 3 and 4 for more detail. The method used in Program 4.4 involves matrix techniques that are readily extended to higher order polynomials as well.

The task here is made more complicated by the fact that most fitting routines, and this one is no exception, apply to a polynomial of the form

$$\Delta\bar{E}_V = a' + b'V + c'V^2 \qquad (4.49)$$

and not to Equation (4.48). Fortunately, provided we 'plot' the $\Delta\bar{E}_V$ values at $(V+\frac{1}{2})$ intervals before fitting to Equation (4.49), one can easily show that

$$a = a', \qquad b = b', \qquad c = c' \qquad (4.50)$$

Once one has obtained the coefficients for Equation (4.49) one must then find the point at which it crosses the V axis. We do this by solving the quadratic Equation (4.49) for $\Delta\bar{E}_V=0$.

$$V_{\mathrm{I}}=\frac{-b'\pm\sqrt{b'^2-4a'c'}}{2a'}$$

Several cases are possible:

(1) $b'^2-4a'c'<0$ no solution

(2) $b'^2-4a'c'=0$ one intersection point

(3) $b'^2-4a'c'>0$ two intersection points

(1) If only a few V levels have been observed it may be that they do not define the curvature well enough over an extended range and the extrapolated Birge–Sponer curve never crosses the V axis.

(3) If two real roots exist, then the V axis is crossed at two places. If one root is negative we can easily reject that on physical grounds. If both are positive we must take the one of lower V—can you see why?

Having determined the intersection point V_{I} we can now find the area under Equation (4.49) by integrating from $V=0$ to $V=V_{\mathrm{I}}$.

$$\text{Area}=\int_0^{V_{\mathrm{I}}} a'+b'V+c'V^2\,\mathrm{d}V=\bar{D}_0$$

$$\bar{D}_0=a'V_{\mathrm{I}}+\frac{b'V_{\mathrm{I}}^2}{2}+\frac{c'V_{\mathrm{I}}^3}{3} \tag{4.51}$$

One is also in a position to provide values for $\overline{\omega}_{\mathrm{e}}$, $\overline{\omega_{\mathrm{e}}x_{\mathrm{e}}}$ and $\overline{\omega_{\mathrm{e}}y_{\mathrm{e}}}$ via Equations (4.48) and (4.50).

$$\overline{\omega_{\mathrm{e}}y_{\mathrm{e}}}=\frac{c'}{3},\qquad \overline{\omega_{\mathrm{e}}x_{\mathrm{e}}}=\frac{c'-b'}{2},\qquad \overline{\omega}_{\mathrm{e}}=\frac{6a'+c'-3b'}{6}$$

If experimentally only a few $\Delta\bar{E}_V$ values have been observed, the extrapolation required is extensive and the predicted value of $\bar{D}_0$ is subject to large uncertainty. Many molecules do not exhibit significant curvature in the Birge–Sponer plot until high V levels are reached, so an extrapolation based on the first few V levels will generally underestimate this contribution. As a result of this the $\bar{D}_0$ value obtained is often much too large and care should be taken, when using automatic extrapolation techniques of this kind, to ensure that the data are of sufficient extent to provide a good basis for extrapolation.

Program 4.4 BIRGE: Birge–Sponer extrapolation

```
90  DIM SX(10),SY(10),A(10,10)
100  REM  BIRGE-SPONER EXTRAPOLATION
110  DIM X(50),Y(50)
120  GOSUB 1000: REM  SET UP DATA
130  GOSUB 4000: REM  SOLVE EQNS
140  PRINT : PRINT "E = ";CA(1);" + ";CA(2);"V + ";CA(3);"V*V"
150  IF CA(3) = 0 THEN XR =  - CA(1) / CA(2): GOTO 220
160 SR = (CA(2) ^ 2 - 4 * CA(3) * CA(1))
170  IF SR < 0 THEN  PRINT "ERROR - DOES NOT HAVE REAL ROOT FOR E=0": GOTO
   310
180 SR =  SQR (SR)
190 XR = ( - CA(2) - SR) / (2 * CA(3))
200  IF XR > 0 THEN  GOTO 220
210 XR = ( - CA(2) + SR) / (2 * CA(3))
220  PRINT "INTERCEPT AT V = ";XR
230 DO = CA(1) * XR + CA(2) * (XR ^ 2) / 2 + CA(3) * (XR ^ 3) / 3
240  PRINT : PRINT "DO = ";DO;" cm-1  (";DO * 1.196E - 2;" kJ mol-1)"
250 CP = CA(3):BP = CA(2):AP = CA(1)
260 WY = CP / 3:WX = (CP - BP) / 2:WE = (6 * AP - 3 * BP + CP) / 6
270  PRINT : PRINT "We = ";WE;"  WeXe = ";WX
280  PRINT "WeYe = ";WY
290 DE = DO + WE / 2 - WX / 4 + WY / 8
300  PRINT : PRINT "De = ";DE;" cm-1  (";DE * 1.196E - 2;" kJ mol-1)"
310  END
1000  REM  MATRIX A
1010  PRINT : INPUT "MOL = ";MN$
1020  PRINT "1 INPUT REAL DATA": PRINT "2 INPUT MOL CONSTANTS"
1030  INPUT "WHICH ? ";DN: IF DN < 1 OR DN > 2 THEN  GOTO 1030
1040  IF DN = 2 THEN  GOTO 1120
1050  PRINT : INPUT "NUMBER OF DATA POINTS = ";NP
1060  PRINT "INPUT V,DE(V) (in cm-1)"
1070  FOR I = 1 TO NP
1080  INPUT V,EV
1090 X(I) = V + .5:Y(I) = EV
1100  NEXT
1110  GOTO 1230
1120 NP = 15: REM  USE FIRST 15 LEVELS
1130  PRINT : INPUT "We = ";WE
1140  INPUT "WeXe = ";WX
1150  INPUT "WeYe = ";WY
1160 W1 = WE - WX + WY:W2 = 2 * WX - 3 * WY:W3 = 3 * WY
1170  FOR V = 0 TO NP - 1
1180 I = V + 1
1190 X(I) = V + .5
1200 Y(I) = W1 - W2 * X(I) + W3 * X(I) * X(I)
1210  PRINT X(I),Y(I)
1220  NEXT
1230  PRINT : PRINT : PRINT
1240  RETURN
4000  REM  SUBROUTINE POLYFIT
4010 NT = 3: REM  THREE TERMS A+BX+CX^2
4020 NM = 2 * NT - 1: REM  NMAX,NTERMS
4030  FOR I = 1 TO NM:SX(I) = 0: NEXT
4040  FOR I = 1 TO NT:SY(I) = 0: NEXT
4050  FOR I = 1 TO NP
4060 X1 = X(I):Y1 = Y(I):XT = 1
4070  FOR J = 1 TO NM:SX(J) = SX(J) + XT:XT = XT * X1: NEXT
4080 YT = Y1
```

```
4090  FOR J = 1 TO NT:SY(J) = SY(J) + YT:YT = YT * X1: NEXT
4100  NEXT : REM  I
4110  FOR J = 1 TO NT: FOR K = 1 TO NT:A(J,K) = SX(J + K - 1): NEXT : NEXT
4120 N = NT: GOSUB 5000:DD = DT: REM  DET(A) IN DD
4130  IF DD <  > 0 THEN  GOTO 4150
4140  FOR J = 1 TO NT:CA(J) = 0: NEXT : GOTO 4200
4150  FOR ZI = 1 TO NT: FOR ZJ = 1 TO NT: FOR ZK = 1 TO NT
4160 A(ZJ,ZK) = SX(ZJ + ZK - 1)
4170  NEXT ZK:A(ZJ,ZI) = SY(ZJ): NEXT ZJ
4180  GOSUB 5000:CA(ZI) = DT / DD
4190  NEXT ZI
4200  RETURN
5000  REM  EVALUATE DET(A)
5010 DT = 1
5020  FOR K = 1 TO N
5030  IF A(K,K) <  > 0 THEN  GOTO 5130
5040  FOR J = K TO N: REM  SWAP COLUMNS
5050 SW = 0: REM  SWAPS FLAG
5060  IF A(J,K) = 0 THEN  GOTO 5110
5070  FOR I = K TO N:ZZ = A(I,J)
5080 A(I,J) = A(I,K):A(I,K) = ZZ
5090  NEXT I
5100 DT =  - DT:SW = 1:J = N
5110  NEXT J
5120  IF SW = 0 THEN DT = 0:K = N: GOTO 5190
5130 DT = DT * A(K,K)
5140  IF (K - N) >  = 0 THEN  GOTO 5190
5150 K1 = K + 1
5160  FOR I = K1 TO N: FOR J = K1 TO N
5170 A(I,J) = A(I,J) - A(I,K) * A(K,J) / A(K,K)
5180  NEXT J: NEXT I
5190  NEXT K
5200  RETURN
```

```
MOL = I2
1 INPUT REAL DATA
2 INPUT MOL CONSTANTS
WHICH ? 1

NUMBER OF DATA POINTS = 9
INPUT V,DE(V) (in cm-1)
?0,213
?10,201
?20,187
?30,172
?40,154
?50,135
?60,111
?70,82
?80,52

E = 212.039281 + -.859502168V + -.0138744589V*V
INTERCEPT AT V = 96.4702823

DO = 12303.8312 cm-1  (147.153821 kJ mol-1)

We = 212.46672  WeXe = .422813855
WeYe = -4.62481962E-03

De = 12409.9583 cm-1  (148.423101 kJ mol-1)
```

```
MOL = P2
1 INPUT REAL DATA
2 INPUT MOL CONSTANTS
WHICH ? 2

We = 780.43
WeXe = 2.804
WeYe = -0.00533
.5              774.804678
1.5             769.148707
2.5             763.460758
3.5             757.740828
4.5             751.988917
5.5             746.205028
6.5             740.389158
7.5             734.541308
8.5             728.661478
9.5             722.749667
10.5            716.805878
11.5            710.830108
12.5            704.822358
13.5            698.782627
14.5            692.710918

E = 777.620661 + -5.62398783V + -.0159901407V*V
INTERCEPT AT V = 106.201033

DO = 44484.2375 cm-1  (532.03148 kJ mol-1)

We = 780.42999  WeXe = 2.80399884
WeYe = -5.33004691E-03

De = 44873.7508 cm-1  (536.69006 kJ mol-1)
```

Program notes

(1) The program operates in two modes. The first inputs experimental vibrational data in the form $V, \Delta\bar{E}_V$, where V refers to the *lower* vibrational level and the energy spacing is in cm^{-1}. In the second mode the vibrational constants $\overline{\omega_e}$, $\overline{\omega_e x_e}$ and $\overline{\omega_e y_e}$ (in cm^{-1}) are used to calculate the vibrational levels.

(2) In the second mode the fitting operation is carried out on the calculated data even though this is unnecessary since the coefficients can be found directly from Equation (4.48). The original constants are then recalculated from the fitting coefficients. This acts as a good test on the accuracy of the program and the 'polyfit' subroutine.

(3) The number of levels output in mode two is set in line 1120, this can be altered if required.

(4) The order of the polynomial to be fitted to is given by NT in line 4010. This subroutine can easily be incorporated into other programs, although higher order polynomials will require redimensioning of the SX, SY and A arrays.

(5) The values calculated for $\bar{D}_0$ and $\bar{D}_e$ in run two for P_2 are substantially above the true values; which illustrates the danger of using vibrational constants for extended extrapolation.

4.3 Vibration and rotation

Up till now we have considered the rotational and vibrational motions of the molecule separately. In reality both motions occur simultaneously, a fact which has considerable significance for infra-red spectra.

4.3.1 *Rigid rotor—harmonic oscillator model*

To a first approximation we can continue to treat the motions as independent, in which case (see Section 2.2) the total energy can be written as

$$\bar{E} = \bar{E}_V + \bar{E}_R \qquad (4.52)$$

where $\bar{E}_V$ and $\bar{E}_R$ are given by Equations (4.19) and (3.11) respectively. The energy of a *ro-vibrational* level is then given by

$$\bar{E}_{V,J} = \bar{\omega}(V + \tfrac{1}{2}) + \bar{B}J(J+1) \qquad (4.53)$$

Since vibrational spacings are much larger than rotational ones, Equation (4.53) means that each vibrational level is accompanied by a set of rotational levels, as shown in *Figure 4.5*.

In fact, the separation between V levels is $\simeq 1000$ times that between J levels, so the diagram is not to scale.

When a transition occurs in infra-red absorption it will be from a given initial (V'', J'') state to some final (V', J') state. The previous selection rules still apply to each part of the transition.

$$\Delta V = \pm 1, \qquad \Delta J = \pm 1 \qquad (4.54)$$

Strictly speaking we can also have $\Delta V = 0$ which represents the pure rotational spectrum of Chapter 3. In an absorption experiment we have $\Delta V = +1$ but the rotational change can now be *either* $\Delta J = +1$ *or* $\Delta J = -1$, since both represent net absorption. For $\Delta J = +1$ we have

$$\Delta\bar{E} = (V + \tfrac{3}{2})\bar{\omega} + \bar{B}(J+1)(J+2) - (V + \tfrac{1}{2})\bar{\omega} - \bar{B}J(J+1)$$

$$\Delta\bar{E} = \bar{\omega} + 2\bar{B}(J+1) \qquad (J = 0, 1, 2\ldots) \qquad (4.55)$$

where J refers to the rotational level in the lower vibrational state. For

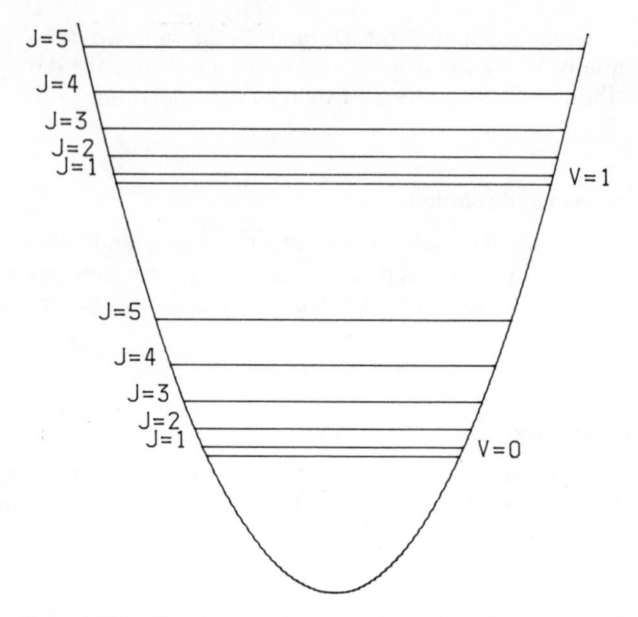

Figure 4.5 The vibration–rotation energy levels for a simple harmonic oscillator–rigid rotor

$\Delta J = -1$ we obtain

$$\Delta\bar{E} = (V + \tfrac{3}{2})\bar{\omega} + \bar{B}(J-1)J - (V + \tfrac{1}{2})\bar{\omega} - \bar{B}J(J+1)$$

$$\Delta\bar{E} = \bar{\omega} - 2\bar{B}J \qquad (J = 1, 2, 3 \ldots) \tag{4.56}$$

where J again refers to the lower vibrational level. The consequences of Equations (4.55) and (4.56) are best seen by reference to *Figure 4.6*.

The $\Delta J = -1$ transitions are labelled *P branch*, whilst those with $\Delta J = +1$ are called the *R branch*. There can also be, under exceptional circumstances, transitions with $\Delta J = 0$ which yield a *Q branch* in which all lines lie at $\bar{\omega}$. The Q branch only arises if there is non-zero *electronic* angular momentum about the internuclear axis such as in the molecule NO.

In general then, we expect a *vibrational* transition to have two band structures associated with the *rotational* motions of the diatomic molecule. A gap (the 'Q branch') will normally occur at the pure vibrational frequency $\bar{\omega}$, usually called the *band origin*. The infra-red vibration-rotation spectrum of HBr is shown in *Figure 4.7*.

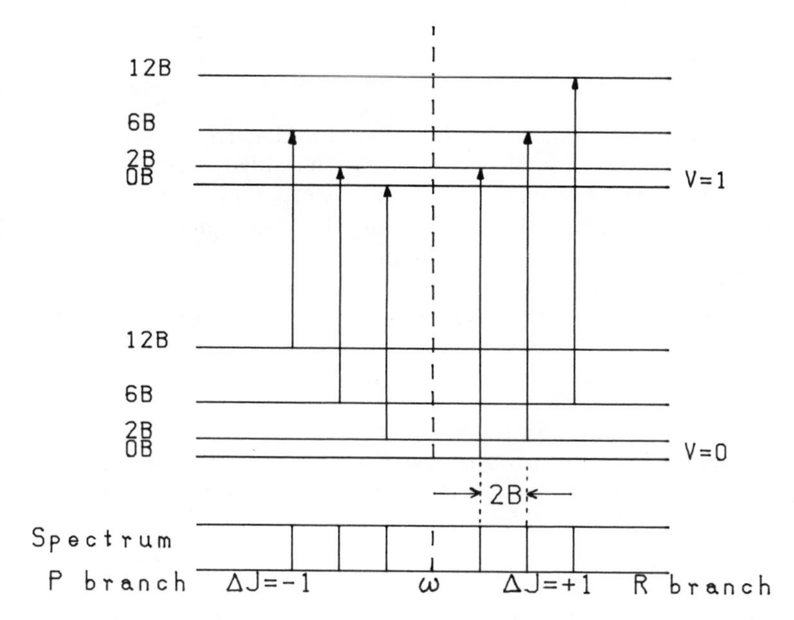

Figure 4.6 Vibration–rotation transitions in a rigid rotor–harmonic oscillator

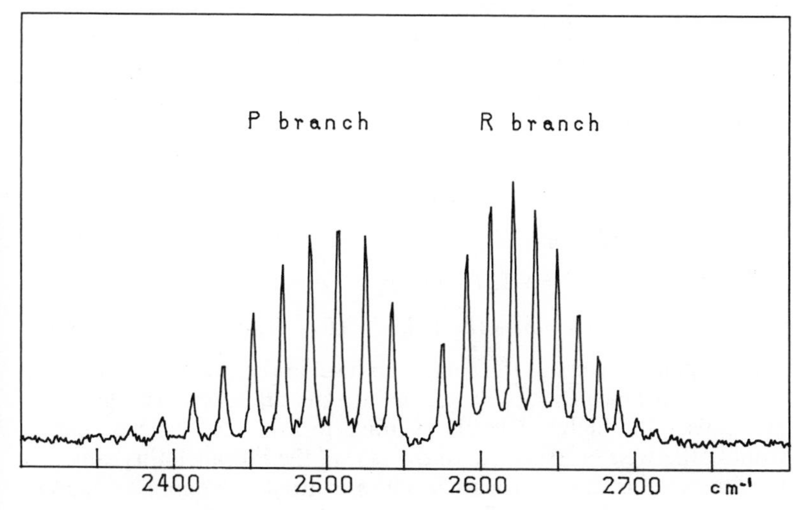

Figure 4.7 The rotational structure accompanying the $V=0$ to $V=1$ transition of gaseous HBr

4.3.2 *The anharmonic rotor*

A careful look at *Figure 4.7* will reveal that the spectral lines are not equidistant but, at high J levels, they converge in the R branch and diverge in the P branch. This is due to vibration–rotation coupling and arises from two distinct sources; the anharmonicity of the potential energy curve and centrifugal distortion (see Section 3.1.6).

The effect of anharmonicity is to move the equilibrium bond length to larger values as V increases. This increases the moment of inertia of the molecule and hence decreases the rotational constant $\bar{B}$. In this case we must use a different value of $\bar{B}$ for each vibrational level. The change in $\bar{B}$ can be represented approximately by

$$\bar{B}_V = \bar{B}_e - \bar{\alpha}_e(V + \tfrac{1}{2}) \tag{4.57}$$

where $\bar{\alpha}_e$ is a small positive constant compared to $\bar{B}_e$.

The centrifugal distortion is treated in the same way as described in Section 3.1.6, except that now the centrifugal distortion constant also depends on the V level.

$$\bar{D}_V = \bar{D}_{eq} - \bar{\beta}_e(V + \tfrac{1}{2}) \tag{4.58}$$

The rotational energy level of a given vibrational level can now be written as

$$\bar{E}_{V,J} = \bar{E}_V + \bar{B}_V J(J+1) - \bar{D}_V J^2(J+1)^2 \tag{4.59}$$

Since most infra-red spectra are taken in absorption at room temperature we find that $V=0$ has the predominant population. A good description of the infra-red spectrum is obtained by considering only the $V=1 \leftarrow 0$ transition and neglecting the centrifugal term in Equation (4.59).

For $\Delta J = +1$ (R branch) we have

$$\Delta\bar{E}_R = \bar{\omega}_V + \bar{B}_1(J+1)(J+2) - \bar{B}_0 J(J+1)$$

$$\Delta\bar{E}_R = \bar{\omega}_V + 2\bar{B}_1 + (3\bar{B}_1 - \bar{B}_0)J + (\bar{B}_1 - \bar{B}_0)J^2 \tag{4.60}$$

and similarly for $\Delta J = -1$ (P branch) we obtain

$$\Delta\bar{E}_P = \bar{\omega}_V - (\bar{B}_1 + \bar{B}_0)J + (\bar{B}_1 - \bar{B}_0)J^2 \tag{4.61}$$

where, again, J refers to the rotational level in the lower vibrational state ($V=0$). Since $\bar{B}_1 < \bar{B}_0$ we see that the quadratic term is negative and causes the higher J levels to move to lower frequencies than would otherwise be expected. In the case of the P branch this causes the lines to spread out at high J, but in contrast it compresses lines together in the R branch. This is exactly the behaviour observed in *Figure 4.7*. One can obtain accurate experimental values for $\bar{B}_0$ and

$\bar{B}_1$ by measuring the spacings between corresponding lines in the P and R branches

$$\Delta\bar{E}_R(J'')-\Delta\bar{E}_P(J'')=2\bar{B}_1(2J''+1)$$
$$\Delta\bar{E}_R(J'')-\Delta\bar{E}_P(J''+2)=2\bar{B}_0(2J''+3) \tag{4.62}$$

In the first case a plot of the energy difference against J'' will have a gradient of $4\bar{B}_1$, whilst in the second we obtain $4\bar{B}_0$. Finally, application of Equation (4.57) to $V=0$ and $V=1$ allows us to find $\bar{\alpha}_e$ and $\bar{\beta}_e$

$$\bar{\alpha}_e=\bar{B}_0-\bar{B}_1, \qquad \bar{\beta}_e=\frac{3\bar{B}_0-\bar{B}_1}{2} \tag{4.53}$$

We can now calculate the bond lengths r_0, r_1 and r_e from the values for $\bar{B}_0$, $\bar{B}_1$, and $\bar{B}_e$ using Equation (3.10). *Table 4.4* lists the values obtained for some isotopic variants of H_2.

Table 4.4 The effects of anharmonicity in some isotopic variants of H_2

Molecule	μ/amu	$\bar{\beta}_e$	$\bar{\alpha}_e$	r_e	r_0	r_1
H_2	0.5041	60.809	2.993	0.7416	0.7509	0.7706
HD	0.6719	45.655	1.9928	0.7413	0.7495	0.7668
HT	0.7556	40.575	1.6705	0.7415	0.7493	0.7655
D_2	1.0074	30.429	1.0492	0.7416	0.7480	0.7643
T_2	1.5085	20.324	0.5922	0.7415	0.7470	0.7583

The figures in *Table 4.4* clearly reveal the effects of anharmonicity, both horizontally and vertically. If one looks at the values for a single molecule it can be seen that the bond lengths increase on moving out of the bottom of the potential well (r_e) to $V=1$ (r_1). The Born–Oppenheimer approximation states that the potential energy curve only depends on the nuclear charges not the masses, in which case all isotopic variants should share an identical curve and we would expect all variants to have the same r_e value (the bottom of the well), as indeed they do. In contrast the heights of the V levels above the well minimum are given by Equation (4.17) (or more accurately Equation (4.29)). Since this vibrational spacing is proportional to $\mu^{-\frac{1}{2}}$ the values of r_0 and r_1 should decrease down the series, which is again correct.

4.4 Polyatomic vibrations

The calculation of the vibrational spectra of polyatomic molecules is a difficult and complex task involving considerable computation and sophisticated mathematical techniques. In this section we examine the basic theory and apply it to the simple example of a linear triatomic molecule. Molecular vibration is one atomic problem that can be solved with almost complete accuracy using classical mechanics only. Indeed we have already seen that for the diatomic molecule the classical vibrational frequency and the quantum mechanical one are identical. In the treatment that follows we shall confine ourselves to small amplitude vibrations for which simple harmonic motion is an accurate description.

4.4.1 *Degrees of freedom and normal modes*

A molecule containing n atoms requires $3n$ coordinates to specify the positions of the atoms. The system is said to have $3n$ *degrees of freedom.* These degrees of freedom can be divided up between translation, rotation and vibration of the molecule. We require 3 degrees of freedom to describe the translation of the centre of mass of the molecule, which leaves $3n-3$ to divide between rotation and vibration. Linear molecules require a further two degrees of freedom to describe rotation (rotation about the internuclear axis has no effect) while non-linear molecules have three rotational degrees of freedom. We thus find that linear molecules possess $3n-5$ vibrational degrees of freedom and non-linear molecules have $3n-6$.

The vibrations themselves are divided into two different classes: bond stretching and bond bending (or angle bending). For an acyclic molecule there are $n-1$ bonds and hence $n-1$ bond stretching vibrations. This leaves $2n-4$ (linear) or $2n-5$ (non-linear) bending vibrations. If the molecule contains any rings then the division between stretching and bending modes depends on the actual geometry. It should be stressed at this point that none of the above implies that a given bond stretching mode involves only two atoms of the molecule or that an angle bending mode only involves three atoms. In general all atoms of the molecule will be involved in each vibrational motion. It is the fact that we are dealing with the simultaneous coupled motions of n atoms that makes the treatment of molecular vibrations so difficult.

In general the cartesian coordinates x, y, z are not the most convenient in which to solve the vibrational equations of motion. From a physical point of view coordinates based on bond stretches and bond bends would seem a lot better. In fact, we can always find a set of coordinates which produce the maximum simplification of the resulting equations of motions, these are called the *normal coordinates* of the system. The vibrational motions along these coordinates are called the *normal modes* of vibration and they have the property that *displacement of the system along the normal coordinates leads to a motion in which all particles move in phase and with the same frequency. Further, if Hooke's law is obeyed, the motion is simple harmonic.*

The classical treatment of molecular vibrations reduces to finding the normal modes and calculating their vibrational frequencies. *Figure 4.8* shows the normal modes for a symmetric linear molecule such as CO_2 and a bent molecule such as H_2O.

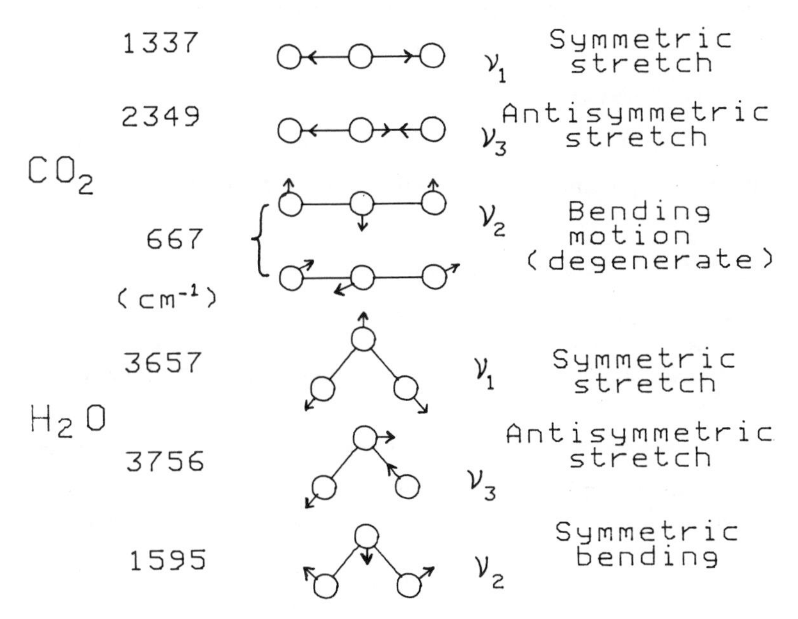

Figure 4.8 The normal modes of CO_2 and H_2O

The terms symmetric and antisymmetric used in *Figure 4.8* refer to how the motion responds to being reflected in a mirror plane passing vertically through the central atom. Such a reflection corresponds to swapping over the two extremum atoms and their motions. If the motion of the molecule looks unaltered, the vibration is symmetric, if it is changed it is antisymmetric. More complex symmetry classifications are possible for many molecules but it is outside the scope of this book to pursue this approach.

The principal advantage of casting the motions in terms of normal modes is that the resulting Schrödinger equation splits naturally into $3n-6$ (or $3n-5$) independent equations, each one associated with a different vibration and identical in form to Equation (4.16). The quantum solutions for each normal mode are the harmonic wavefunctions (4.20) with an energy level spacing given by Equation (4.17). The same selection rule, $\Delta V = \pm 1$, also applies. We must of course also examine the vibrational motions themselves to determine whether they produce a changing dipole moment, as discussed in Section 2.6. We thus find that ν_1 of CO_2 is infra-red inactive but ν_2 and ν_3 both produce vibrational spectra. We shall return to this point again in Chapter 5. The *total* vibrational wavefunction is a product of the individual wavefunctions for each normal mode,

$$\psi_V = \psi_1(Q_1)\psi_2(Q_2)\ldots\psi_{3n-6}(Q_{3n-6}) \tag{4.64}$$

and the energy is given by

$$E_V = E_1 + E_2 + E_3 \ldots E_{3n-6} \tag{4.65}$$

where Q_i represents the motions of the atoms associated with normal mode i.

4.4.2 *The calculation of vibrational frequencies*

The classical mechanics of coupled oscillators is best treated in Lagrange's formulation of Newton's laws of motion, since it only requires expressions for the kinetic and potential energy of the system. The Lagrangian is defined by

$$L = T - V \tag{4.66}$$

It is also convenient to define the kinetic and potential energies in terms of *displacements* of particles from their equilibrium positions.

$$T = \frac{1}{2}\sum_{i=1}^{3n} m_i(\Delta\dot{x}_i)^2 \tag{4.67}$$

where Δx_i represents changes in x, y, z for each particle, i.e. Δx_i ($i = 1, 2, 3$) corresponding to Δx, Δy and Δz for particle 1, and $m_i = m_1$ ($i = 1, 2, 3$) and so on. We also now introduce *mass weighted* coordinates, q, such that

$$q_i = \sqrt{m_i}\,\Delta x_i \tag{4.68}$$

then Equation (4.67) becomes

$$2T = \sum_i^{3n} \dot{q}_i^2 \tag{4.69}$$

If we expand the potential function in much the same way that we did for Equation (4.28) and retain only the harmonic terms we obtain

$$2V = \sum_{ij}\left(\frac{\partial^2 V}{\partial q_i\,\partial q_j}\right)_0 \cdot q_i q_j \tag{4.70}$$

It must be stressed here that the q_i in Equation (4.70) represents motion of individual particles—not changes in bond length as in Equation (4.28). We can again identify the second order partial derivative with a *force constant*:

$$f_{ij}=\left(\frac{\partial^2 V}{\partial q_i\,\partial q_j}\right)_0 \qquad \text{and} \qquad f_{ij}=f_{ji} \tag{4.71}$$

One is now in a position to solve Lagrange's equations of motion for each particle, these are

$$\frac{\mathrm{d}}{\mathrm{d}t}\left(\frac{\partial T}{\partial \dot{q}_j}\right)+\left(\frac{\partial V}{\partial q_j}\right)=0 \tag{4.72}$$

which leads to $3n$ simultaneous *differential* equations of the form

$$\ddot{q}_j+\sum_i^{3n} f_{ij}q_i=0 \qquad j=1\ldots 3n \tag{4.73}$$

These equations are the classical equations of motion for coupled harmonic oscillators and have a general solution

$$q_i=A_i\cos(\lambda^{\frac{1}{2}}t+\phi) \qquad \text{and frequency} \qquad \nu=\lambda^{\frac{1}{2}}/2\pi \tag{4.74}$$

substitution of Equation (4.74) into (4.73) yields $3n$ *algebraic* equations

$$\sum_{i=1}^{3n}(f_{ij}-\delta_{ij}\lambda)A_i=0 \qquad j=1\ldots 3n \tag{4.75}$$

in which δ_{ij} is the Kronecker delta with value $\delta_{ij}=1$ if $i=j$, and 0 otherwise. These have non-trivial solutions only if the determinant of the coefficients is zero

$$\begin{vmatrix} f_{11}-\lambda & f_{12} & f_{13} & \cdots & f_{1,3n} \\ f_{21} & f_{22}-\lambda & f_{23} & \cdots & f_{2,3n} \\ \cdots & \cdots & \cdots & \cdots & \cdots \\ f_{3n,1} & f_{3n,2} & f_{3n,3} & \cdots & f_{3n,3n}-\lambda \end{vmatrix}=0 \tag{4.76}$$

The solution of this determinant by direct expansion and solving for the roots of the resulting polynomial in λ^{3n} is not practicable except for small n. Instead we now turn to a matrix representation of the problem.

Equation (4.76) is mathematically equivalent to finding the eigenvalues of the symmetric matrix **F**, where **F** is defined as

$$\mathbf{F}=\begin{pmatrix} f_{11} & f_{12} & \cdots & f_{1,3n} \\ f_{21} & f_{22} & \cdots & f_{2,3n} \\ \cdots & \cdots & \cdots & \cdots \\ f_{3n,1} & f_{3n,2} & \cdots & f_{3n,3n} \end{pmatrix} \tag{4.77}$$

An eigenvalue problem can be defined as finding the matrix **L** such that

$$\mathbf{FL}=\mathbf{L\Lambda} \tag{4.78}$$

where $\mathbf{\Lambda}$ is a diagonal matrix made up of the solutions (roots) of the secular determinant (4.76).

$$\mathbf{\Lambda}=\begin{pmatrix} \lambda_1 & 0 & 0 & 0 \\ 0 & \lambda_2 & 0 & 0 \\ 0 & 0 & \lambda_3 & 0 \\ \cdots & \cdots & \cdots & \cdots \\ 0 & 0 & 0 & \lambda_{3n} \end{pmatrix}$$

The matrix **L** is called the eigenvector matrix and $\mathbf{\Lambda}$ the eigenvalue matrix. A property of **L** is that it is orthogonal, i.e. $\tilde{\mathbf{L}}\mathbf{L}=\mathbf{I}$ where **I** is the unit matrix and $\tilde{}$ represents a matrix transpose. From this fact it follows from Equation (4.78) that

$$\tilde{\mathbf{L}}\mathbf{FL}=\tilde{\mathbf{L}}\mathbf{L\Lambda}=\mathbf{\Lambda} \tag{4.79}$$

We say that the matrix **L** *diagonalizes* the matrix **F**. The columns of **L** are just the eigenvectors associated with the eigenvalues in $\mathbf{\Lambda}$, i.e. column 1 is the eigenvector for λ_1, and so on. A more detailed analysis shows that the matrix $\tilde{\mathbf{L}}$ is just the matrix that transforms the mass weighted coordinates **q** into normal modes **Q**

$$\mathbf{Q}=\tilde{\mathbf{L}}\mathbf{q}=\tilde{\mathbf{L}}\mathbf{mx} \tag{4.80}$$

where **m** is a $3n\times 3n$ diagonal matrix with elements $\sqrt{m_i}$. The eigenvalues of the *force constant matrix* **F** thus yield the vibrational frequencies via Equation (4.74), and the normal modes of the system via **L**.

4.4.3 *The normal modes of a linear triatomic molecule*

As a simple example we consider the vibrational motions of a linear triatomic molecule. One can separate the motions into those along the internuclear axis and those perpendicular to it (see *Figure 4.8*).

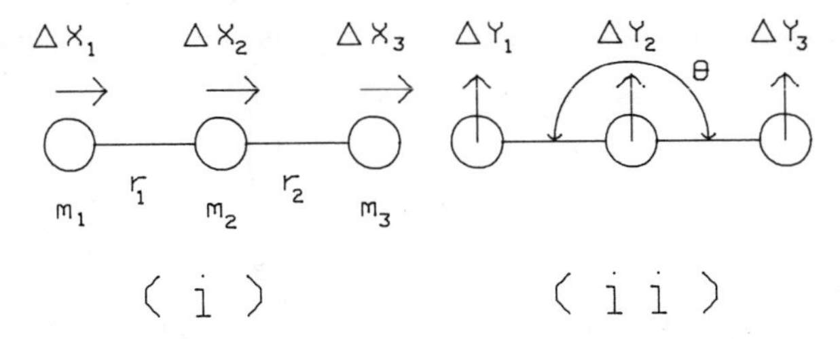

Figure 4.9 Coordinates for determining the vibrational motions of a linear triatomic molecule; (i) bond stretching, (ii) bond bending

This enables one to solve the problem in two independent parts, as demonstrated in *Figure 4.9*.

(i) Bond stretches

In order to evaluate Equation (4.70), one must have an expression for the potential energy, and we use here the so called *valence force* approximation. In this approximation only those atoms connected by bonds can exert forces on each other, further we choose this force to be to simple harmonic.

$$V=\tfrac{1}{2}k_1(\delta r_1)^2+\tfrac{1}{2}k_2(\delta r_2)^2 \tag{4.81}$$

where k is a force constant and $\delta r_1=\Delta x_2-\Delta x_1$, $\delta r_2=\Delta x_3-\Delta x_2$.

We can express the potential function (4.81) in terms of mass weighted coordinate (4.68)

$$V=\tfrac{1}{2}k_1\left(\frac{q_2}{\sqrt{m_2}}-\frac{q_1}{\sqrt{m_1}}\right)^2+\tfrac{1}{2}k_2\left(\frac{q_3}{\sqrt{m_3}}-\frac{q_2}{\sqrt{m_2}}\right)^2 \tag{4.82}$$

It is now a simple task to evaluate the force constant matrix elements using Equation (4.71) to obtain

$$\mathbf{F}=\begin{pmatrix} \dfrac{k_1}{m_1} & \dfrac{-k_1}{\sqrt{m_1m_2}} & 0 \\ \dfrac{-k_1}{\sqrt{m_1m_2}} & \dfrac{k_1+k_2}{m_2} & \dfrac{-k_2}{\sqrt{m_2m_3}} \\ 0 & \dfrac{-k_2}{\sqrt{m_2m_3}} & \dfrac{k_2}{m_3} \end{pmatrix} \tag{4.83}$$

In the simplest possible case of $k_1=k_2=m_1=m_2=m_3=1$ we have

$$\mathbf{F}=\begin{pmatrix} 1 & -1 & 0 \\ -1 & 2 & -1 \\ 0 & -1 & 1 \end{pmatrix}$$

and

$$\mathbf{L}=\begin{pmatrix} 1/\sqrt{3} & -1/\sqrt{6} & -1/\sqrt{2} \\ 1/\sqrt{3} & 2/\sqrt{6} & 0 \\ 1/\sqrt{3} & -1/\sqrt{6} & 1/\sqrt{2} \end{pmatrix} \qquad \mathbf{\Lambda}=\begin{pmatrix} 0 & 0 & 0 \\ 0 & 3 & 0 \\ 0 & 0 & 1 \end{pmatrix}$$

and the normal modes and frequencies are

$$Q_1=\frac{1}{\sqrt{3}}(\Delta x_1+\Delta x_2+\Delta x_3) \qquad \text{freq}=0$$

$$Q_2=\frac{1}{\sqrt{6}}(-\Delta x_1+2\,\Delta x_2-\Delta x_3) \qquad \text{freq}=\sqrt{3}/2\pi$$

$$Q_3=\frac{1}{\sqrt{2}}(-\Delta x_1+\Delta x_3) \qquad \text{freq}=1/2\pi$$

The first of these corresponds to pure translation of the molecule. Q_2 and Q_3 are the antisymmetric and symmetric stretches shown in *Figure 4.8*.

(ii) Bond bends

Since the two bending motions in a linear triatomic molecule are degenerate (identical) we only need to consider one of the pair: in this case the xy plane. Within the valence force approximation we can write the bending motion potential energy as

$$V=\tfrac{1}{2}k_3r_1r_2(\delta\theta)^2 \tag{4.84}$$

The form of Equation (4.84) is slightly different to Equation (4.81) since angle bending force constants are usually quoted in the form $k_3=k_\theta/r_1r_2$ to maintain the same units of Nm^{-1}. The value of k_3 will be much less than k_1 or k_2 since bond bending requires less force than bond stretching. Provided the amplitude of oscillations is small we can approximate $\delta\theta$ by

$$\delta\theta=\frac{\Delta y_1-\Delta y_2}{r_1}+\frac{\Delta y_3-\Delta y_2}{r_2} \tag{4.85}$$

substituting Equation (4.85) into Equation (4.84) and using mass weighted coordinates, we can evaluate the force constant matrix elements as before

$$\mathbf{F}=\begin{pmatrix} \dfrac{k_3}{m_1}\left(\dfrac{r_2}{r_1}\right) & \dfrac{-k_3}{\sqrt{m_1m_2}}\left(1+\dfrac{r_2}{r_1}\right) & \dfrac{k_3}{\sqrt{m_1m_3}} \\ \dfrac{-k_3}{\sqrt{m_1m_2}}\left(1+\dfrac{r_2}{r_1}\right) & \dfrac{k_3}{m_2}\left(\dfrac{r_1}{r_2}+\dfrac{r_2}{r_1}+2\right) & \dfrac{-k_3}{\sqrt{m_2m_3}}\left(1+\dfrac{r_1}{r_2}\right) \\ \dfrac{k_3}{\sqrt{m_1m_3}} & \dfrac{-k_3}{\sqrt{m_2m_3}}\left(1+\dfrac{r_1}{r_2}\right) & \dfrac{k_3}{m_3}\left(\dfrac{r_1}{r_2}\right) \end{pmatrix} \quad (4.86)$$

It can be seen from this that a knowledge of the bond lengths r_1 and r_2 is required to evaluate r_1/r_2, except for symmetrical triatomic molecules where the ratio must be unity. The force constant matrix (4.86) only produces one non-zero eigenvalue, corresponding to one component of the doubly degenerate bending mode.

By splitting the vibrations up into Equations (4.83) and (4.86) we only have to solve a 3×3 eigenvalue problem for each type of motion. A non-linear triatomic molecule would not separate out in such a simple manner and we would have to construct a single 6×6 force constant matrix.

Program 4.5 calculates the vibrational frequencies and normal modes of a linear triatomic molecule using Equations (4.83) and (4.86). The program incorporates a routine for finding the eigenvalues and eigenvectors of a symmetric matrix. More detail on the methodology of determining eigenvectors can be found in Reference 3, although the routine presented here uses the Jacobi method outlined in Reference 5. The program is configured to solve two 3×3 problems separately. This is not very elegant since overall three eigenvalues of

Table 4.5 Vibrational frequencies, force constants and equilibrium bond lengths for several linear triatomic molecules

Molecule	ν_1	ν_3	ν_2	k_1	k_2	k_3/r_1r_2	r_1	r_2
CO_2	1337	2349	667	$\begin{pmatrix}1680\\1420\end{pmatrix}$	—	57	1.16	1.16
CS_2	657	1523	397	$\begin{pmatrix}810\\690\end{pmatrix}$	—	23.4	1.56	1.56
HCN	2089	3312	712	580	1790	20	1.06	1.15
BrCN	580	2187	368	420	1690	17	1.93	1.15
SCO	859	2079	527	800	1420	37	1.54	1.16

Note: ν in cm^{-1}, k in Nm^{-1} and r in Å

zero frequency also have to be found, but it does keep the program complexity to a minimum.

Table 4.5 lists force constants and experimental vibrational frequencies of several triatomic molecules. In the case of symmetrical molecules, $k_1 = k_2$ and both ν_1 and ν_3 can be used to estimate k_1. Unfortunately it is often found that these values do not agree very well and both values are given in the table. The disagreement is a result of the failure of the valence force approximation to describe the true potential energy adequately. More realistic calculations would add more terms to Equations (4.81) and (4.84) to obtain an improved fit (see Problem 4.16).

Program 4.5 NMODES: Linear triatomic normal modes

```
100  REM  LINEAR TRIATOMIC VIBRATIONS
110 N = 3:IM = 12:EPS = 1E - 6:BF = 0
120  DIM T(N,N),S(N,N),TN(N,N),R(N,N),ST(N,N),TS(N,N),RS(N,N),M(N),V(N)
130  GOSUB 2000: REM  SET UP MATRIX
140  PRINT : PRINT "EIGENVALUES BEING CALCULATED..."
150 IT = 1
160  GOSUB 5500: REM  LARGEST OFF DIAG. ELEMENT IN T
170  IF  ABS (MAX) < EPS THEN  GOTO 230
180  GOSUB 5000: REM  ONE ITERATION
190 IT = IT + 1: IF IT < IM THEN  GOTO 160
200  PRINT : PRINT "EPS CONDITION (";EPS;") NOT MET AFTER"
210  PRINT "ITERATION ";IT;"  MAX = ";MAX
220  GOTO 250
230  PRINT : PRINT "EPS CONDITION (";EPS;")"
240  PRINT "MET AT ITERATION ";IT: PRINT
250  GOSUB 3000: REM  RESULTS
260 BF = BF + 1: IF BF = 2 THEN  GOTO 290
270  PRINT : PRINT "BENDING FREQUENCY"
280  GOSUB 2070: GOTO 140
290  END
2000  REM  SET UP T MATRIX
2010  PRINT : INPUT "INPUT K1,K2 (Nm-1) ";K1,K2
2020  PRINT : INPUT "INPUT M1,M2,M3 (amu) ";M(1),M(2),M(3)
2030  PRINT : INPUT "INPUT K3 (Nm-1), R1/R2 ";K3,RB
2040 T(1,1) = K1:T(1,2) =  - K1:T(1,3) = 0
2050 T(2,2) = K1 + K2:T(2,3) =  - K2:T(3,3) = K2
2060  GOTO 2110
2070  REM  BENDING COORDS
2080 T(1,1) = K3 / RB:T(1,2) =  - K3 * (1 + 1 / RB):T(1,3) = K3
2090 T(2,2) = K3 * (RB + 1 / RB + 2):T(2,3) =  - K3 * (1 + RB)
2100 T(3,3) = K3 * RB
2110 T(2,1) = T(1,2):T(3,1) = T(1,3):T(3,2) = T(2,3)
2120  FOR ZI = 1 TO N: FOR ZJ = 1 TO N:T(ZI,ZJ) = T(ZI,ZJ) /  SQR (M(ZI) *
   M(ZJ)): NEXT : NEXT
2130  RETURN
3000  REM  PRINT RESULTS
3010  PRINT : PRINT "EIGEN VALUE MATRIX": PRINT
3020  FOR ZI = 1 TO N: FOR ZJ = 1 TO N
3030  PRINT  TAB( (ZJ - 1) * 12 + 1);
```

```
3040  PRINT  INT ( ABS (T(ZI,ZJ)) * 1E8) / 1E8 *  SGN (T(ZI,ZJ));"   ";: NEXT
    : PRINT : NEXT
3050  PRINT : PRINT : PRINT "EIGEN VECTOR MATRIX": PRINT
3060  FOR ZI = 1 TO N: FOR ZJ = 1 TO N
3070  PRINT  TAB( (ZJ - 1) * 12 + 1);
3080  PRINT  INT ( ABS (R(ZI,ZJ)) * 1E8) / 1E8 *  SGN (R(ZI,ZJ));"   ";: NEXT
    : PRINT : NEXT
3090  PRINT : PRINT
3100  FOR ZI = 1 TO N:V(ZI) =  SQR ( ABS (T(ZI,ZI)))
3110 V(ZI) = V(ZI) * 130.284: REM  cm-1
3120  IF  ABS (V(ZI)) < 2 THEN  GOTO 3180
3130  PRINT "FREQUENCY";ZI;" = "; INT (V(ZI) * 100) / 100;"  cm-1"
3140  FOR ZJ = 1 TO N:DX = R(ZJ,ZI) /  SQR (M(ZJ))
3150  IF BF = 0 THEN  PRINT "DELTAX";: GOTO 3170
3160  PRINT "DELTAY";
3170  PRINT ZJ;" = "DX: NEXT : PRINT
3180  NEXT
3190  RETURN
5000  REM  CALC COS AND SIN
5010 LA =  - T(P,Q):MU = .5 * (T(P,P) - T(Q,Q))
5020 NU =  SQR (LA * LA + MU * MU)
5030 C =  SQR ((NU +  ABS (MU)) / (2 * NU))
5040 SI =  SGN (MU): IF SI = 0 THEN SI = 1
5050 S = SI * LA / (2 * NU * C)
5060  REM  CREATE IDENTITY MATRIX
5070  FOR ZI = I TO N: FOR ZJ = 1 TO N:S(ZI,ZJ) = 0: NEXT :S(ZI,ZI) = 1: NEXT

5080  REM  NOW S
5090 S(P,P) = C:S(Q,Q) = C:S(P,Q) = S:S(Q,P) =  - S
5100  REM  NOW ST
5110  FOR ZI = 1 TO N: FOR ZJ = 1 TO N
5120 ST(ZI,ZJ) = S(ZJ,ZI)
5130  NEXT : NEXT
5140  REM  CALC R MATRIX
5150  IF IT > 1 THEN  GOTO 5180
5160  FOR ZI = 1 TO N: FOR ZJ = 1 TO N:R(ZI,ZJ) = 0
5170  NEXT :R(ZI,ZI) = 1: NEXT
5180  REM  CALC R*S AND PUT INTO R
5190  FOR ZI = 1 TO N: FOR ZJ = 1 TO N:RS(ZI,ZJ) = 0
5200  FOR ZS = 1 TO N:RS(ZI,ZJ) = RS(ZI,ZJ) + R(ZI,ZS) * S(ZS,ZJ): NEXT
5210  NEXT : NEXT
5220  FOR ZI = 1 TO N: FOR ZJ = 1 TO N:R(ZI,ZJ) = RS(ZI,ZJ): NEXT : NEXT

5230  REM  PRODUCT ST*T*S INTO TN
5240  FOR ZI = 1 TO N: FOR ZJ = 1 TO N:TS(ZI,ZJ) = 0
5250  FOR ZS = 1 TO N:TS(ZI,ZJ) = TS(ZI,ZJ) + T(ZI,ZS) * S(ZS,ZJ): NEXT
5260  NEXT : NEXT
5270  FOR ZI = 1 TO N: FOR ZJ = 1 TO N:TN(ZI,ZJ) = 0
5280  FOR ZS = 1 TO N:TN(ZI,ZJ) = TN(ZI,ZJ) + ST(ZI,ZS) * TS(ZS,ZJ): NEXT

5290  NEXT : NEXT
5300  REM  PUT TN INTO T
5310  FOR ZI = 1 TO N: FOR ZJ = 1 TO N:T(ZI,ZJ) = TN(ZI,ZJ): NEXT : NEXT

5320  RETURN
5500  REM  LARGEST OFF DIAG
5510 MAX =  ABS (T(1,2)):P = 1:Q = 2
5520  FOR ZI = 1 TO N: FOR ZJ = 1 TO N: IF ZJ = ZI THEN  GOTO 5540
5530  IF  ABS (T(ZI,ZJ)) > MAX THEN P = ZI:Q = ZJ:MAX =  ABS (T(ZI,ZJ))
5540  NEXT : NEXT
5550  RETURN
```

```
INPUT K1,K2 (Nm-1) 580, 1790

INPUT M1,M2,M3 (amu) 1, 12, 14

INPUT K3 (Nm-1), R1/R2 20, 0.92174

EIGENVALUES BEING CALCULATED...

EPS CONDITION (1E-06)
MET AT ITERATION 8

EIGEN VALUE MATRIX

647.77936    0            2E-08
0            257.577783   -3.5E-07
0            -2.7E-07     4E-08

EIGEN VECTOR MATRIX

.922358S2    .33498913    .19245008
-.37338756   .64508703    .66666666
.09917903    -.68676415   .7200823

FREQUENCY1 = 3315.92  cm-1
DELTAX1 = .922358521
DELTAX2 = -.107787707
DELTAX3 = .0265067115

FREQUENCY2 = 2090.95  cm-1
DELTAX1 = .334989135
DELTAX2 = .186220586
DELTAX3 = -.18354544

BENDING FREQUENCY

EIGENVALUES BEING CALCULATED...

EPS CONDITION (1E-06)
MET AT ITERATION 3

EIGEN VALUE MATRIX

29.6926052   0            0
0            0            0
0            0            0

EIGEN VECTOR MATRIX

.85484352    .48511027    -.18414821
-.47423175   .87445298    .10215779
.21058675    0            .97757517

FREQUENCY1 = 709.92  cm-1
DELTAY1 = .854843529
DELTAY2 = -.136898917
DELTAY3 = .0562816767
```

Program notes

(1) Line 110 sets some important constants for the eigenvector routine. N (N × N) is the matrix size, IM is the maximum number of iterations and EPS is the maximum size permitted for an off-diagonal element in the eigenvalue matrix. If EPS cannot be achieved by IM iterations this fact is displayed and the calculation continues.
(2) The array, T, corresponds initially to **F** but it is converted into **Λ** by the Jacobi routine. Array R corresponds to the eigenvector matrix **L**.
(3) Lines 3000–3090 print out the eigenvalue and eigenvector matrices. Lines 3100–3190 print out the non-zero frequencies and the cartesian displacements associated with that normal mode. The eigenvalues are converted to cm^{-1} in line 3110, where $\nu = (1000L)^{\frac{1}{2}}/(200\pi c)\lambda^{\frac{1}{2}}$.

The approach adopted here to calculate the normal modes and vibrational frequencies is conceptually simple but rather clumsy in practice for several reasons.

(i) The calculations use cartesian displacements of the atoms whereas internal coordinates based on bond stretching and bond bending would be more physical.
(ii) The force constant matrix is also in terms of cartesian coordinates, but again the forces are most readily expressed in terms of bond stretches and bond bends.
(iii) The calculation includes zero eigenvalues corresponding to translations and rotations, it would be more efficient to remove these at the outset.
(iv) No account is taken of symmetry—a factor which can often simplify the problem considerably.

Very elegant and powerful matrix techniques have been developed to overcome these shortcomings but the theoretical formalism is somewhat more complex. The interested reader is referred to the books by Wilson *et al.* and Califano, cited at the end of this chapter, for more detail. Excellent introductions to these techniques can also be found in Barrow and Steinfeld.

4.5 References

1. Tennant-Smith, J., *BASIC Statistics*, Butterworths (1985).
2. Morse, P.M., *Phys. Rev.*, **34**, 57 (1929).
3. Mason, J.C., *BASIC Matrix Methods*, Butterworths (1984).
4. Bevington, P.R., *Data Reduction and Error Analysis for the Physical Sciences*, McGraw-Hill (1969).
5. Ralston, A., *A First Course in Numerical Analysis*, McGraw-Hill (1965).

4.6 Further reading

Barrow, G.M., *Introduction to Molecular Spectroscopy*, McGraw-Hill (1962).
Califano, S., *Vibrational States*, John Wiley & Sons (1976).
Herzberg, G., *Spectra of Diatomic Molecules*, Van Nostrand (1950).
Herzberg, G., *Infra-red and Raman Spectra*, Van Nostrand (1945).
Steinfeld, J.I., *Molecules and Radiation*, The MIT Press (1974).
Wilson, E.B., Decius, J.C. and Cross, P.C., *Molecular Vibrations*, McGraw-Hill (1955).

PROBLEMS

(4.1) Show that Equation (4.15) does indeed follow from Equation (4.14).
(4.2) Use Equations (4.20) to (4.22) to construct the vibrational wavefunctions ψ_3 and ψ_4.
(4.3) The Hermite polynomials can be generated from the relationship $y.H_n(y)=n.H_{n-1}(y)+0.5.H_{n+1}(y)$ given that $H_0(y)=1$ and $H_1(y)=2y$. Modify Program 4.1 to calculate the vibrational wavefunctions for any V.
(4.4) Write a program using the trapezium rule to evaluate the integrals (4.25) numerically. Use this to confirm the selection rule $\Delta V=\pm 1$, then use the relationship in Problem 4.3 to prove it analytically.
(4.5) Write a program that, given the masses and (harmonic) vibrational frequency for a diatomic molecule, outputs the force constant and Boltzmann population for the first $5V$ levels.
(4.6) The fundamental and first overtone transitions of CO occur at $2143.29\,\mathrm{cm}^{-1}$ and $4259.65\,\mathrm{cm}^{-1}$. Evaluate $\bar{\omega}$, $\overline{\omega_e x_e}$, the zero point energy, force constant and $\bar{D}_0$. Comment on the calculated value for $\bar{D}_0$ compared with the actual value of 11.09 eV.
(4.7) Modify Program 4.3 to output the first 30 levels and calculate the energy levels expected for HCl (*Table 4.3*). Why do the anharmonic spacings become negative?
(4.8) Write a program that calculates the Morse potential (4.35) given $\bar{\omega}_e$, $\bar{D}_e$ and the masses. Use this program to plot actual potential curves for H_2, HCl, HI and N_2 using the data in *Table 4.3* and the following information

	H_2	HCl	HI	N_2
$\bar{D}_e/\mathrm{cm}^{-1}$	38289	37232	25807	79892
$r_e/\mathrm{Å}$	0.7416	1.2746	1.604	1.094

(4.9) Draw the anharmonic energy levels on the curves produced in

Problem 4.8. Also calculate the $\overline{\omega_e x_e}$ values expected. Comment on any discrepancies between these values and those in *Table 4.3*.

(4.10) Verify that Equation (4.45) results from substitution of Equation (4.41) into (4.44). Hint:

$$\sum_0^z (n+1) = \tfrac{1}{2}(z+1)(z+2), \qquad \sum_0^z 1 = (z+1)$$

(4.11) Write a program, analogous to Program 3.2, which calculates the wavelengths and intensities of the P and R branch transitions accompanying a vibrational transition of a diatomic molecule within the harmonic oscillator–rigid rotor approximation. Now extend this to include anharmonicity.

(4.12) The *rotational* structure accompanying vibrational transitions in linear triatomic molecules is a little more complex than in diatomic molecules. One must divide the vibrational motions into those that are *parallel* to the internuclear axis (ν_1 and ν_3) and those that are *perpendicular* to it (ν_2). The selection rules now become

Parallel	$\Delta V = \pm 1$	$\Delta J = \pm 1$	(PR branches)
Perpendicular	$\Delta V = \pm 1$	$\Delta J = 0, \pm 1$	(PQR branches)

Modify the program of Problem 4.11 to include these possibilities. Unfortunately the picture is complicated further by *nuclear spin*, a point examined in Chapter 5.

(4.13) How many normal modes do the following molecules have: HBr, HCN, BF_3 and XeF_6?

(4.14) Use Program 4.5 and the data in *Table 4.5* to determine the normal modes of HCN and BrCN. Make a drawing of each molecule using the correct bond lengths and attach arrows showing the accurate relative displacements of each atom.

(4.15) The total vibrational energy of a polyatomic molecule depends on the V levels occupied in *each* mode. We can represent a particular total energy level by the shorthand notation (V_1, V_2, V_3). A level with two quanta in ν_2 would thus be written as (0, 2, 0). Assign the transitions in water (*Figure 4.8*) that are responsible for the following lines, given that the lower state is (0, 0, 0) in all cases: 1595, 3152, 3657, 3756, 5331, 6872, 7252 cm^{-1}. Can you think of an algorithm to do this efficiently on a computer?

(4.16) For a symmetric linear triatomic molecule like CO_2 the potential energy expression (4.81) only involves *one* force constant, k_1. In this case we can improve the potential by including a term $0.5k_{13}(\delta r_{13})^2$ as well. Determine the new force constant matrix that results from this addition and modify Program 4.5 to use it. Find values for the two force constants for CO_2 and CS_2 using the frequencies given in *Table 4.5*.

Chapter 5

Raman and electronic spectra

ESSENTIAL THEORY

In this final chapter we turn to the visible/UV region of the spectrum. The excitation of electronic motions, corresponding to transitions between different *electronic* states, gives rise to absorption and emission in this region. We shall, however, leave discussion of this until later in the chapter. Instead we shall first address ourselves to a process that arises from *photon scattering* rather than photon absorption—the so-called *Raman* effect.

Raman scattering occurs in all regions of the spectrum but its intensity is proportional to the fourth power of the incident light frequency. Hence the Raman effect is predominantly observed in visible, rather than infra-red or microwave, spectroscopy.

5.1 The Raman effect

If a parallel monochromatic light beam is incident on a transparent sample then a small percentage of the light is scattered in all directions. The majority of the scattered light has the same frequency as the original light and an intensity proportional to the fourth power of the light frequency. This process, called *Rayleigh scattering*, is the reason why the sky appears blue—since the blue portion of white light is scattered more strongly than the red portion.

A closer examination reveals that about 0.1% of the scattered light is scattered at *different frequencies* to the original, both higher and lower. It is this phenomenon that is called the Raman effect. The weakness of the Raman effect makes it difficult to observe using conventional light sources but the advent of high-power lasers has been responsible for a considerable renaissance in this technique and it is now an invaluable spectroscopic tool.

It is found that the shift in frequencies is independent of the incident light frequency and is thus a property of the molecule rather than the light itself. The lines at lower frequency than the original light are called *Stokes lines* and those at higher frequency are called *Anti-*

Stokes lines. This terminology is borrowed from the description of fluorescence even though the two processes are fundamentally different.

The Raman effect arises from inelastic scattering of photons in which the outgoing photon gains or loses a set amount of energy, equal to a change of vibrational or rotational quantum state in the molecule. In order to understand how this occurs we must examine the effect that an external electric field has on the charge distribution of the molecule.

5.1.1 *The classical Raman effect*

If a molecule is subjected to an electric field $\vec{E}$ it induces a small electric dipole moment $\vec{\mu}$ in the molecule as the electrons move towards the positive end of the field and the nuclei move in the opposite direction. We can write this as

$$\vec{\mu} = \dot{\alpha}\vec{E} \tag{5.1}$$

or more fully as

$$\begin{pmatrix} \mu_x \\ \mu_y \\ \mu_z \end{pmatrix} = \begin{pmatrix} \alpha_{xx} & \alpha_{xy} & \alpha_{xz} \\ \alpha_{yx} & \alpha_{yy} & \alpha_{yz} \\ \alpha_{zx} & \alpha_{zy} & \alpha_{zz} \end{pmatrix} \begin{pmatrix} E_x \\ E_y \\ E_z \end{pmatrix} \tag{5.2}$$

The symmetric matrix α is called the *molecular polarizability* and has six independent components. The polarizability describes how readily the electron charge distribution around the nuclei can be distorted by the external electric field. The theoretical calculation of these components is a formidable task and it requires large-scale computation to obtain accurate results. Fortunately, symmetry considerations are sufficient to determine the selection rules that apply to Raman scattering and a detailed knowledge of the polarizability components will not be required.

In order to determine the consequences of Equation (5.2), it is sufficient to consider just one representative component and to assume that we can describe the electric field of the light by Equation (2.28), one obtains

$$\mu = \alpha E_0 \cos 2\pi\nu t \tag{5.3}$$

The oscillating dipole obtained has the same frequency as the light itself and is responsible for the dominant Rayleigh scattering. However, the polarizability components themselves are affected by molecular motions such as vibration and rotation since these alter the electron distribution that the incoming photon encounters. One thus expects vibrations and rotations to produce oscillating polarizability components and indeed this is the case. In the case of *vibration* this can be represented by

$$\alpha=\alpha_0+\beta \cos 2\pi\nu_{osc}t \qquad (\beta \ll \alpha_0) \tag{5.4}$$

and Equation (5.3) then becomes

$$\begin{aligned}\mu &= \{\alpha_0+\beta \cos 2\pi\nu_{osc}t\} . E_0 \cos 2\pi\nu t \\ &= \alpha_0 E_0 \cos 2\pi\nu t+\tfrac{1}{2}\beta E_0\{\cos 2\pi(\nu+\nu_{osc})t+\cos 2\pi(\nu-\nu_{osc})t\}\end{aligned} \tag{5.5}$$

The dipole moment of Equation (5.5) will give rise to radiation at the original light frequency (Rayleigh scattering) from the first term, and at higher and lower frequencies (Raman scattering) from the other two terms. We thus see that for a motion to be Raman active it must produce a changing polarizability component.

5.1.2 *The quantum Raman effect*

Equation (5.5) states that classically we would expect two lines of similar intensity either side of the Rayleigh line. In fact this is not the case experimentally and we must turn to a quantum description in order to understand why. We will not pursue the mathematical treatment here but *Figure 5.1* provides a reasonable pictorial representation of the Raman process. We can image it as being absorption and re-emission involving an intermediate 'virtual' state, although it must be stressed that no real intermediate state is involved.

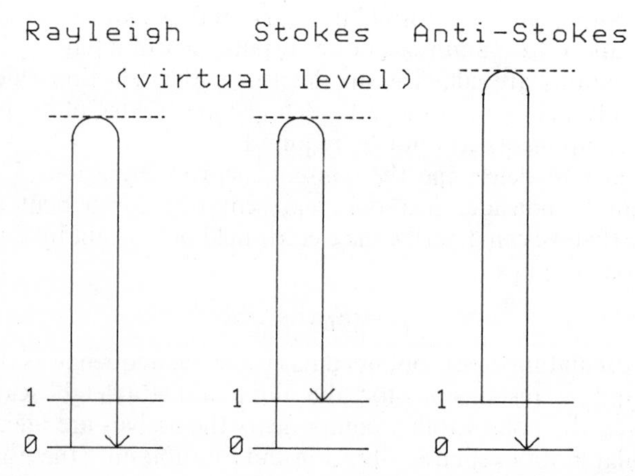

Figure 5.1 A simple representation of Rayleigh and Raman scattering

It can be seen from *Figure 5.1* that the Stokes and Anti-Stokes lines start from different initial levels. To a first approximation the intensities of the lines will be proportional to the Boltzmann populations of the initial state. In the case of a vibrating diatomic molecule at room temperature only the $V=0$ and $V=1$ levels have significant populations. Application of Equation (2.25) yields

$$\frac{I_{\text{Stokes}}}{I_{\text{Anti-Stokes}}}=e^{-(E_1-E_0)/kT}=e^{-h\nu_{\text{osc}}/kT} \qquad (5.6)$$

where E_1 and E_0 refer to $V=1$ and $V=0$ respectively. The Anti-Stokes lines are thus usually very much weaker than the Stokes lines. The lines will occur at $\nu_0 \pm (\omega_e - 2\omega_e x_e)$ where ν_0 is the incident light frequency.

The selection rules for Raman activity require that two criteria be met

(i) There must be a changing polarizability component

(ii) vibration: $\Delta V = \pm 1$ (5.7)

rotation: $\Delta J = 0, \ \pm 2$

The $\Delta J = -2$ transitions are labelled O branches and the $\Delta J = +2$ are labelled S branches in an extension of the P, Q, R nomenclature of Section 4.3.

5.2 Pure rotational Raman spectra—heteronuclear diatomic molecules

In the case of a pure rotational Raman spectrum of a molecule like HCl we find a strong line at the Q branch corresponding to $\Delta J=0$ (Rayleigh scattering). The positions of the Raman lines can be found from Equations (5.7) and (3.11) to be

$$\nu_0 \pm 2B(2J+3) \qquad J=0, 1, 2 \ldots \qquad (5.8)$$

To be strictly consistent with our previous notation, the pure rotational spectrum can only arise from $\Delta J = +2$ since ΔJ is defined as $J' - J''$ where J' refers to the upper state and J'' the lower one. There are, however, *two* series of lines since the lower J'' state can be the initial or final state in the scattering. We say that the Stokes and Anti-Stokes lines are both S branches.

The first line of each branch occurs at a distance of $6B$ from the Q line and subsequent lines occur at intervals of $4B$, as illustrated in *Fig. 5.2*.

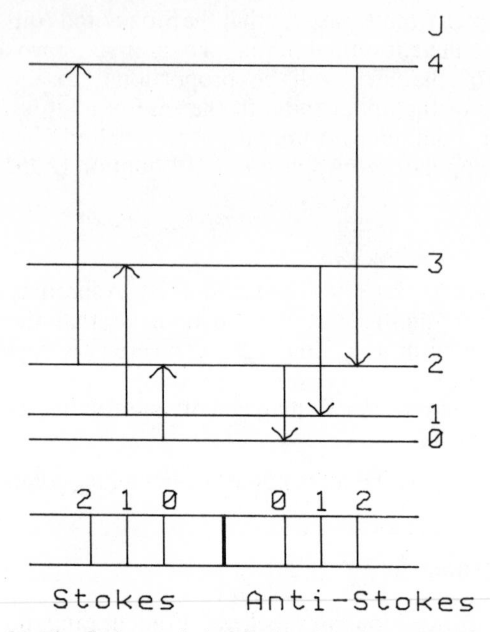

Figure 5.2 The Stokes and Anti-Stokes lines in rotational Raman scattering. J'' is the initial level in the Stokes line and the final one in the Anti-Stokes line

The intensities of the lines will be proportional to the population of the initial rotational level. A detailed analysis reveals that there is a (complicated) J dependency as well but the dominant effect is the Boltzmann population. Program 5.1 calculates the rotational Raman spectrum of *heteronuclear* diatomic molecules on this basis.

Program 5.1 ROTRAM: Diatomic rotational Raman spectrum

```
100  DIM ET(50),IA(50),IS(50)
110  REM   ROTATIONAL RAMAN SPECTRA
120  GOSUB 1000: REM  GET DATA
130  GOSUB 2000: REM  INTENSITIES
140  END
1000  REM  DATA INPUT
1010  PRINT : PRINT "DIATOMIC DATA": PRINT
1020  PRINT : INPUT "MOLECULE = ";MN$
1030  PRINT : INPUT "Be and Alphae = ";BE,AE
1040 BO = BE - AE / 2: REM  ROT CONSTANT FOR V=0
1050  PRINT : INPUT "TEMPERATURE (K) = ";T
1060  PRINT : INPUT "MAXIMUM LEVEL (<50) = ";NL
1070 KT = 0.6950817 * T: REM  kT IN cm-1
1080  RETURN
```

```
2000  REM  CALC INTENSITIES
2010  FOR J = 0 TO NL
2020 ES = J * (J + 1) * B0: REM  STARTING LEVEL FOR STOKES LINE
2030 EA = (J + 2) * (J + 3) * B0: REM  ANTI-STOKES LEVEL
2040 ET = 2 * B0 * (2 * J + 3): REM  ENERGY DIFFERENCE
2050 IS(J) = (2 * J + 1) *  EXP ( - ES / KT)
2060 IA(J) = (2 * J + 5) *  EXP ( - EA / KT)
2070 ET(J) =  INT (ET * 100 + .5) / 100: REM  ROUND RESULT
2080  NEXT
2090  REM  NORMALIZE INTENSITIES
2100  PRINT : PRINT "ROTATIONAL RAMAN INTENSITIES"
2110 M1 = 0
2120  FOR J = 0 TO NL
2130  IF IA(J) > M1 THEN M1 = IA(J)
2140  IF IS(J) > M1 THEN M1 = IS(J)
2150  NEXT
2160  FOR J = 0 TO NL
2170 IA(J) = IA(J) / M1:IS(J) = IS(J) / M1
2180 IA(J) =  INT (IA(J) * 10000 + 0.5) / 10000
2190 IS(J) =  INT (IS(J) * 10000 + 0.5) / 10000
2200  NEXT
2210  PRINT : PRINT "|DELTAE|/cm-1","STOKES","ANTI-STOKES"
2220  FOR J = 0 TO NL
2230  PRINT ET(J),IS(J),IA(J)
2240  NEXT
2250  RETURN
```

```
DIATOMIC DATA

MOLECULE = CO

Be and Alphae = 1.9313, 0.01748

TEMPERATURE (K) = 300

MAXIMUM LEVEL (<50) = 20

ROTATIONAL RAMAN INTENSITIES

|DELTAE|/cm-1   STOKES          ANTI-STOKES
11.54           .1117           .5285
19.23           .329            .7001
26.92           .5285           .8362
34.61           .7001           .932
42.3            .8362           .9861
49.99           .932            1
57.68           .9861           .9779
65.37           1               .9258
73.06           .9779           .851
80.75           .9258           .7609
88.44           .851            .6629
96.13           .7609           .5633
103.82          .6629           .4674
111.51          .5633           .3789
119.2           .4674           .3003
126.89          .3789           .2328
134.58          .3003           .1766
142.27          .2328           .1311
149.96          .1766           .0953
157.65          .1311           .0679
165.34          .0953           .0473
```

5.3 Rotational level populations, symmetry and nuclear spin

In our treatment of rotational intensities in Chapter 3 we were able to avoid a point of considerable complexity—the problem of calculating the rotational populations of molecules with a centre of symmetry, such as O_2, CO_2 or C_2H_4. This was possible because such molecules do not possess permanent dipole moments and, hence, do not exhibit rotational spectra. This is not the case for Raman scattering where the requirement is that there be an oscillating polarizability component and we must now apply ourselves to the above task.

In order to gain the advantage of clarity we shall restrict the discussion here to homonuclear diatomic molecules with a closed electronic configuration, i.e. a Σ state. This is not as restricting as may first appear since most diatomic molecules have a closed configuration in their ground electronic state.

We write the total wavefunction (Sections 2.2 and 2.4) as

$$\psi = \psi_e \psi_v \psi_r \tag{5.9}$$

The total wavefunction must either change sign or stay the same upon reflection of *all* particle coordinates through the origin ($x \rightarrow -x$, $y \rightarrow -y$, $z \rightarrow -z$). This must be the case since such an inversion cannot change the energy of the molecule and the energy itself depends only on the magnitude of the wavefunction, not its sign. Whether a sign change occurs depends on how ψ_e and ψ_r behave under the inversion (ψ_v is always positive). An investigation of the rotational wavefunctions shows that $J=0, 2, 4 \ldots$ are *positive* (+), but $J=1, 3, 5 \ldots$ are *negative* (−). The electronic wavefunction can be + (Σ^+) or − (Σ^-). States are said to have *even parity* if they are +, and *odd parity* if they are −.

The total sign change is found by multiplying together the signs of ψ_e and ψ_r. In this way if $\psi_e = \Sigma^+$ then $J=0, 2, 4 \ldots$ are + and $J=1, 3, 5 \ldots$ are −. However if $\psi_e = \Sigma^-$ then $J=0, 2, 4 \ldots$ are − and $J=1, 3, 5 \ldots$ are +.

Two further symmetry classifications are also possible. If we invert the *electron* coordinates only then the *electronic* wavefunction either stays the same (gerade) or changes sign (ungerade). This classification depends only on the nuclear charge, not the mass, hence $^{16}O^{16}O$ and $^{16}O^{18}O$ both possess the same g and u classifications. The final symmetry property depends on how the *nuclear* wavefunction behaves upon nuclei exchange. The wavefunction is either symmetric (s) or antisymmetric (a) to this exchange. This last symmetry label only applies if both nuclei are *identical*, thus $^{16}O^{18}O$ does not possess this classification but $^{16}O^{16}O$ does.

A detailed analysis reveals that not all combinations of these symmetry properties are allowed. *Figure 5.3* shows the allowed combinations for Σ states.

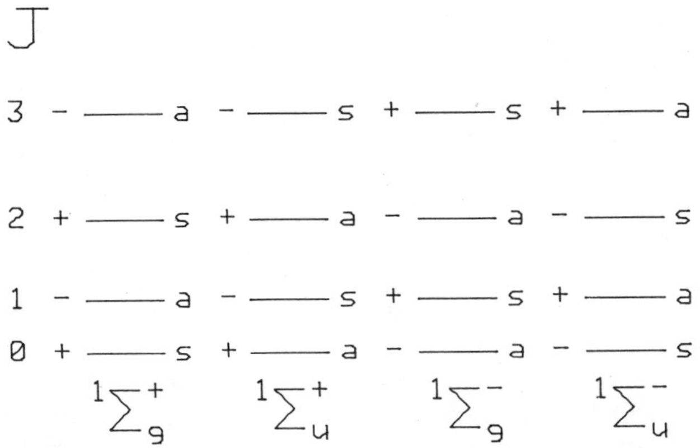

J				
3	− —— a	− —— s	+ —— s	+ —— a
2	+ —— s	+ —— a	− —— a	− —— s
1	− —— a	− —— s	+ —— s	+ —— a
0	+ —— s	+ —— a	− —— a	− —— s
	$^1\Sigma_g^+$	$^1\Sigma_u^+$	$^1\Sigma_g^-$	$^1\Sigma_u^-$

Figure 5.3 The rotational level parities and symmetries of homonuclear diatomic molecules in Σ electronic states. For heteronuclear molecules the *g/u* and *a/s* labels must be omitted

The same classifications also apply to linear polyatomic molecules possessing a centre of symmetry such as CO_2 or C_2H_2.

The reason that these classifications are important is that nuclei themselves possess intrinsic spin—rather like electrons. If the wavefunction for nuclear spin is included in Equation (5.9) it is found that this term produces a different *statistical weight* or degeneracy for the symmetric and antisymmetric levels. This manifests itself in the relative populations of the rotational levels and hence in the observed line intensities. In the case of a homonuclear diatomic molecule in which each nucleus has spin I one finds

$$\frac{n_s}{n_a}=\frac{I+1}{I} \quad (I \text{ is integral or zero})$$
$$\frac{n_s}{n_a}=\frac{I}{I+1} \quad (I \text{ is half integral}) \qquad (5.10)$$

The difference arises because nuclei with integral spin obey Bose–Einstein statistics, but nuclei with half integral spins (and electrons) follow Fermi–Dirac statistics. *Table 5.1* lists nuclear spins for several common elements and isotopes.

In the case of a molecule such as O_2 ($I=0$) this leads to a *complete absence* of antisymmetric levels. The ground electronic state of O_2 is $^3\Sigma_g^-$ which has the same parity properties as the $^1\Sigma_g^-$ state of *Figure 5.3*. We thus find that all the *even* numbered rotational levels are completely missing. A consequence of this is that every alternate line in the spectrum is absent—a point discussed more fully in Problem

Table 5.1 Nuclear spins of some commonly occurring nuclei

H	$\frac{1}{2}$	^{14}N	1	^{31}P	1/2
D	1	^{15}N	$\frac{1}{2}$	^{35}Cl	3/2
T	$\frac{1}{2}$	^{16}O	0	^{37}Cl	3/2
^{12}C	0	^{17}O	$\frac{1}{2}$	^{79}Br	3/2
^{13}C	$\frac{1}{2}$	^{18}O	0	^{81}Br	3/2
^{14}C	0	^{19}F	$\frac{1}{2}$	^{127}I	5/2

5.2. In fact, many nuclear spins were first assigned using line intensities from rotational spectra.

The power of this symmetry exclusion is demonstrated by the fact that $^{16}O^{18}O$, which has no *s*, *a* classification, shows all lines in the spectrum. In the case of H_2 ($I=\frac{1}{2}$) we have a ratio of symmetric level to antisymmetric levels of 1:3. Since the ground electronic state is $^1\Sigma_g^+$, we find that the even J levels have a statistical weighting of 1 and the odd J levels have a statistical weighting of 3. This leads to an intensity alternation in the observed spectrum. We must not forget that the rotational levels also have a $(2J+1)$ degeneracy term, so the total degeneracy of the even levels is $(2J+1)$ whilst that of the odd levels is $3\times(2J+1)$.

Exactly the same considerations apply to linear polyatomic molecules containing a centre of symmetry and possessing only two identical atoms with non-zero spin. Thus, CO_2 behaves like O_2 (but this time the *odd* levels are missing since the ground electronic state is $^1\Sigma_g^+$, and C_2H_2 exhibits the 1:3 intensity alternation that H_2 does.

Program 5.2 calculates the statistical weighting and total degeneracies of the rotational levels for a general linear polyatomic molecule with a centre of symmetry in a ground Σ electronic state. It includes generalizations of Equation (5.10) and the rules illustrated in *Figure 5.3* to handle several nuclei of different spins.

Program 5.2 NUSTAT: Linear molecule nuclear statistics

```
90  DIM SI(10),SY$(10),AM(10)
100  REM  LINEAR MOLECULE NUCLEAR SPIN STATISTICS
110  GOSUB 1000: REM  GET DATA
120  GOSUB 2000: REM  DETERMINE WEIGHTS
130  GOSUB 3000: REM  PRINT RESULTS
140  END
1000  REM  INPUT DATA
1010  PRINT : INPUT "NUMBER OF ATOMS = ";NA
1020 NH =  INT (NA / 2):NC = NA - NH * 2: REM  NUMBER IN ONE HALF AND CEN
   TRE
1030  REM  INPUT SYMBOL,MASS,SPIN FOR EACH ATOM
1040  FOR I = 1 TO NH
1050  PRINT "ATOM";I;: INPUT "   SYMBOL,MASS,SPIN ";SY$(I),AM(I),SI(I)
```

```
1060  NEXT
1070  IF NC = 0 THEN  GOTO 1090
1080  PRINT : INPUT "CENTRAL ATOM  SYMBOL,MASS,SPIN ";CA$,CM,CS
1090  PRINT : INPUT "ELECTRONIC PARITY (+/-) ";EP$
1100  IF EP$ <  > "+" AND EP$ <  > "-" THEN  GOTO 1090
1110  RETURN
2000  REM  DETERMINE STATISTICS TYPE
2010 ST = 1: REM  INITIALISE STATISTICS
2020  FOR I = 1 TO NH
2030  IF SI(I) = 0 THEN  GOTO 2050
2040  IF  ABS (SI(I) / 2 -  INT (SI(I)) / 2) > 0.01 THEN ST =  - ST
2050  NEXT
2060 ST$ = "B": IF ST =  - 1 THEN ST$ = "F"
2070  REM  EVALUATE NUCLEAR SPIN VALUES
2080 S1 = 1:S2 = 1
2090  FOR I = 1 TO NH
2100 S1 = S1 * ((2 * SI(I) + 1) ^ 2)
2110 S2 = S2 * (2 * SI(I) + 1)
2120  NEXT
2130 NA = (S1 - S2) / 2:NS = (S1 + S2) / 2: REM  ANTISYM AND SYM VALUES
2140  IF NC = 0 THEN  GOTO 2160
2150 NA = NA * (2 * CS + 1):NS = NS * (2 * CS + 1)
2160  REM  NOW ASSIGN TO J LEVELS
2170  IF EP$ = "+" AND ST$ = "F" THEN NE = NA:NO = NS
2180  IF EP$ = "-" AND ST$ = "F" THEN NE = NS:NO = NA
2190  IF EP$ = "+" AND ST$ = "B" THEN NE = NS:NO = NA
2200  IF EP$ = "-" AND ST$ = "B" THEN NE = NA:NO = NS
2210  RETURN
3000  REM  OUTPUT RESULTS
3010  FOR I = 1 TO NH:M$ = M$ + SY$(I): NEXT
3020  IF NC = 0 THEN  GOTO 3040
3030 M$ = M$ + CA$
3040  FOR I = NH TO 1 STEP  - 1:M$ = M$ + SY$(I): NEXT
3050  PRINT : PRINT : PRINT "MOLECULE ";M$;" OBEYS ";
3060  IF ST$ = "F" THEN  PRINT "FERMI STATISTICS "
3070  IF ST$ = "B" THEN  PRINT "BOSE STATISTICS "
3080  PRINT : PRINT "ROTATIONAL AND SPIN DEGENERACIES"
3090  PRINT "J   STAT.WT.    ROT.DEG.    TOTAL DEGEN."
3100  FOR J = 0 TO 3 STEP 2
3110 RE = 2 * J + 1:RO = 2 * J + 3
3120  PRINT J;"       ";NE,RE,RE * NE
3130  PRINT J + 1;"       ";NO,RO,RO * NO
3140  NEXT
3150  RETURN

NUMBER OF ATOMS = 3
ATOM1   SYMBOL,MASS,SPIN O,16,0

CENTRAL ATOM  SYMBOL,MASS,SPIN C,12,0

ELECTRONIC PARITY (+/-) +

MOLECULE OCO OBEYS BOSE STATISTICS

ROTATIONAL AND SPIN DEGENERACIES
J   STAT.WT.    ROT.DEG.    TOTAL DEGEN.
0      1          1              1
1      0          3              0
2      1          5              5
3      0          7              0
```

```
NUMBER OF ATOMS = 5
ATOM1    SYMBOL,MASS,SPIN H,1,0.5
ATOM2    SYMBOL,MASS,SPIN N,14,1

CENTRAL ATOM  SYMBOL,MASS,SPIN C,12,0

ELECTRONIC PARITY (+/-) +


MOLECULE HNCNH OBEYS FERMI STATISTICS

ROTATIONAL AND SPIN DEGENERACIES
J    STAT.WT.     ROT.DEG.     TOTAL DEGEN.
0       15        1              15
1       21        3              63
2       15        5              75
3       21        7              147
```

Program notes

(1) Only half the molecule is input starting from the left-hand end.
(2) The masses are not used, but they serve the purpose of showing which isotope is being used (see also Problem 5.5).
(3) Lines 2010–2060 determine whether the statistics are Fermi or Bose overall.
(4) Nuclear spin degeneracies are calculated in lines 2080–2150.
(5) The tests at 2170–2200 assign the spin wavefunctions to even and odd rotational levels.

5.4 The vibrational Raman effect and the polarizability ellipsoid

In order to see how condition (i) of Equation (5.7) applies to vibrational motions we must examine the polarizability in more detail. In general the polarizability of a molecule is anisotropic—that is the electron distribution is more easily distorted in some directions than others. This fact is most readily seen if $1/\sqrt{\alpha}$ is plotted in various directions from the electrical centre of the molecule (often the centre of mass too). It can be shown that the surface obtained is an ellipsoid, called the *polarizability ellipsoid*. This is defined by three principal semi-axes in much the same way that the moment of inertia has three principal axes. Indeed, in many cases the momental and polarizability axes are the same. Since the polarizability ellipsoid is inversely proportional to $\sqrt{\alpha}$ it has its smallest value in the most polarizable direction and its largest value in the least polarizable direction. *Figure 5.4* shows the changing polarizability ellipsoid accompanying vibration of a diatomic molecule.

Since this changing polarizability is independent of whether the molecule is homonuclear or not we find that *all* diatomic molecules

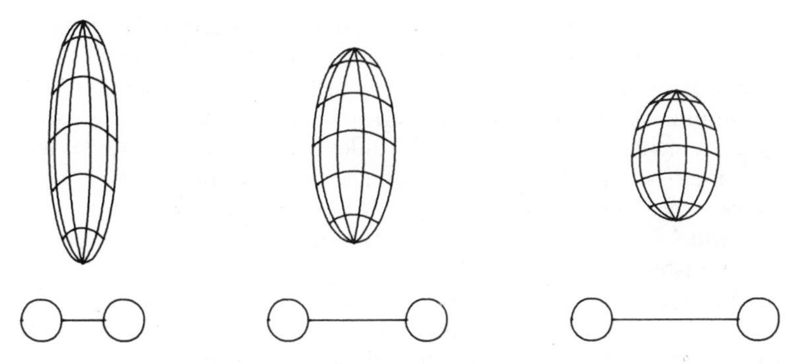

Figure 5.4 The changing polarizability ellipsoid accompanying vibration of a diatomic molecule

exhibit a vibrational Raman effect. This is very different from the case for infra-red absorption in which a fluctuating dipole moment is required—thus rendering all homonuclear diatomics infra-red inactive.

We can extend this approach to the motion of polyatomic vibrations, although here we find we must proceed with considerable caution if we are not to be misled. *Figure 5.5* shows polarizability ellipsoids for the vibrational modes of CO_2. Also included are curves of α *vs* ξ, where ξ represents the generalized displacement from equilibrium for each mode. The condition of a fluctuating polarizability requires that $(\partial\alpha/\partial\xi)_0 \neq 0$.

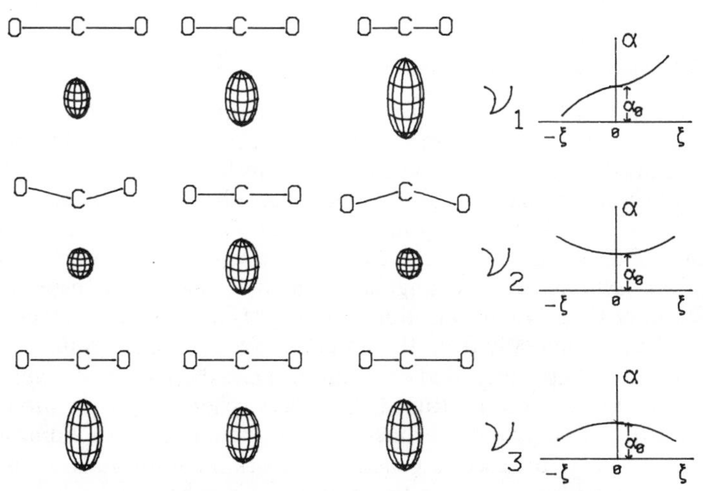

Figure 5.5 The polarizability ellipsoids for the vibrational modes of CO_2

At first glance it would appear that all three modes possess changing polarizabilities and hence would be Raman active. In fact, ν_2 and ν_3 are symmetrical in ξ and hence $(\partial\alpha/\partial\xi)_0 = 0$ since there is a minimum (ν_2), or a maximum (ν_3), in the plot of α *vs* ξ at zero displacement. Thus, for small amplitude oscillations, these modes will be Raman *inactive*. On the other hand, ν_1 will be Raman *active*. This is exactly the opposite to the situation found in infra-red absorption (Section 4.4.1) where we saw that ν_1 is infra-red inactive but ν_2 and ν_3 are infra-red active.

In fact it is a general rule that *for a molecule possessing a centre of symmetry, the active Raman and infra-red modes are mutually exclusive*. For molecules that do not possess a centre of symmetry, modes may be both Raman and infra-red active. Thus it is found that BF_3, which is planar and symmetrical, exhibits the expected exclusion in its four fundamental modes of vibration (3 IR active and 1 Raman active), whereas H_2O has all three modes present in both its infra-red and Raman spectra.

5.5 Electronic spectra of diatomic molecules

The high resolution electronic spectroscopy of polyatomic molecules is an immensely complex topic and beyond the scope of an introductory text of this type. We can, however, use diatomic molecules to illustrate many of the features associated with the description of electronic transitions whilst maintaining some degree of simplicity.

5.5.1 *Diatomic electronic states*

In an ideal world the electronic states of a diatomic molecule would be found by solving the appropriate Schrödinger equation. In fact, such calculations are extremely difficult, especially for heavier atoms and it is only recently that reliable potentials have become available from *ab-initio* calculations for some molecules. From a more practical point of view it is sufficient to classify an electronic state in terms of various symmetries and angular momenta, since it is these that determine the allowed transitions. The energy of the state can then be found experimentally from the observed electronic spectrum.

The most frequently used description is based on a *vector coupling* scheme, rather like the Russell–Saunders scheme used for atomic states. In this method the one-electron orbitals corresponding to solutions of H_2^+ are used as a basic framework. These orbitals are then populated with electrons up to the required number for the diatomic molecule under consideration. In such a scheme we must explicitly apply the *Pauli principle*—which forbids any two electrons to share all the same quantum numbers.

We can represent the total wavefunction ψ as a product of the individual molecular orbitals ϕ.

$$\psi = \phi_1(q_1)\phi_2(q_2)\phi_3(q_3)\ldots \tag{5.11}$$

where q_i represents the coordinates of electron i. The total angular momentum of state ψ (in units of $h/2\pi$) is found from

$$\Lambda = \left|\sum_i \lambda_i\right| \tag{5.12}$$

where λ is the angular momentum about the internuclear axis associated with an *individual* orbital ($\sigma = 0$, $\pi = 1$, $\delta = 2\ldots$). The *total* angular momentum is labelled in an analogous manner; $\Lambda = 0, 1, 2$ is denoted by Σ, Π, Δ. The energy is found to depend only on $|\Lambda|$ so that Π, $\Delta\ldots$ states are doubly degenerate—corresponding to the angular momentum arising from clockwise or anticlockwise motion of the electrons about the internuclear axis. Σ states are non-degenerate.

The spins of the electrons add vectorially according to

$$\vec{S} = \sum_i \vec{s}_i \tag{5.13}$$

The spin state corresponding to the quantum number S is $(2S+1)$ degenerate. This degeneracy, called the *multiplicity* of the state, is given the term singlet, doublet, triplet for $2S+1$ values of 1, 2, 3, etc.

The symmetry classifications outlined in Section 5.3 are also used to distinguish between the electronic states. In the case of Σ states the electronic parity (+ or −) is appended to the description. This is superfluous in all other cases since the degenerate electronic states are made up of one of each parity ($\Pi^+\Pi^-$, etc.). The states are also given their g or u symmetry where applicable. The *overall* g or u symmetry is readily found by multiplying together the g and u symmetries of the individual orbitals using the relationships: $g \times g = g$, $u \times u = g$, $u \times g = u$. The resultant electronic state is denoted by a *term symbol* in the following manner

$$^{2S+1}|\Lambda|^{+/-}_{g/u} \tag{5.14}$$

No examples of the coupling procedure and the systematic application of the Pauli principle will be given here. The interested reader should consult Reference 1 for more detail.

Some electronic configurations denoted by Equation (5.11) lead to more than one electronic state or term symbol. This happens when there is more than one way of arranging the electrons (including spins)

Table 5.2 Electronic states arising from different electronic configurations of C_2

Electronic configuration	*States*
KK $(2s\sigma_g)^2(2s\sigma_u)^2(2p\pi_u)^4$	$^1\Sigma_g^+$
KK $(2s\sigma_g)^2(2s\sigma_u)^2(2p\pi_u)^3(2p\sigma_g)$	$^3\Pi_u$, $^1\Pi_u$
KK $(2s\sigma_g)^2(2s\sigma_u)^2(2p\pi_u)^2(2p\sigma_g)^2$	$^3\Sigma_g^-$, $^1\Delta_g$, $^1\Sigma_g^+$
KK $(2s\sigma_g)^2(2s\sigma_u)(2p\pi_u)^4(2p\sigma_g)$	$^3\Sigma_u^+$, $^1\Sigma_u^+$
KK $(2s\sigma_g)^2(2s\sigma_u)(2p\pi_u)^3(2p\sigma_g)^2$	$^3\Pi_g$, $^1\Pi_g$

within the same configuration. *Table 5.2* gives the electronic states arising from various configurations of C_2.

The selection rules governing electronic transitions in diatomic molecules are found to be

$$\Delta\Lambda = 0, \pm 1 \qquad \Delta S = 0$$

$$\Sigma^+ \leftrightarrow \Sigma^+ \qquad \Sigma^- \leftrightarrow \Sigma^- \qquad \Sigma^+ \not\leftrightarrow \Sigma^- \tag{5.15}$$

$$g \leftrightarrow u \qquad g \not\leftrightarrow g \qquad u \not\leftrightarrow u$$

where $\leftrightarrow$ means that the transition is allowed and $\not\leftrightarrow$ means that it is forbidden. Thus we see that $^1\Sigma_u^+ \leftarrow {}^1\Sigma_g^+$ and $1\Pi \leftarrow {}^1\Sigma^-$ are allowed but $^3\Sigma_u^+ \leftarrow {}^1\Sigma_g^+$ and $^1\Delta \leftarrow {}^1\Sigma^-$ are forbidden.

5.6 Electronic, vibrational and rotational energy levels

An electronic transition does not consist of a single line but rather a large number of closely spaced lines. We divide the lines up into two types: *coarse* and *fine* depending on the relative spacing. The coarse lines arise from changes in vibrational state accompanying the electronic transition. Each coarse line has associated with it fine lines arising from changes in rotational state. At high resolution an electronic spectrum may contain hundreds or thousands of lines!

Provided that we can write the wavefunction in the form (5.9), then the total energy will be given by

$$E = E_e + E_v + E_r \tag{5.16}$$

in units of cm^{-1} this is commonly written

$$T = T_e + G + F$$

where G and F are given by Equations (4.47) and (4.59), although generally cubic terms in G and centrifugal distortion terms in F are not included

$$G = \overline{\omega}_e(V + \tfrac{1}{2}) - \overline{\omega_e x_e}\,(V + \tfrac{1}{2})^2 \tag{5.17}$$

$$F = \bar{B}_V J(J+1) \tag{5.18}$$

Provided we know the selection rules governing ΔV and ΔJ, we can use Equations (5.17) and (5.18) to calculate the energy differences for the transitions $(T_e', V', J') \leftarrow (T_e'', V'', J'')$.

5.7 Vibrational structure and the Franck–Condon principle

The selection rules that govern vibrational transitions in the infra-red are not applicable to electronic transitions. Those rules arise from the fact that both initial and final vibrational state belong to the *same* electronic state. When this is no longer the case then *all possible changes* in vibrational levels may occur. This is not to say that all transitions occur with equal probability—they do not—but there is no simple selection rule to indicate the favoured transitions.

In order to calculate the intensities of the vibrational transitions we must return to the basic definition of the transition probability (2.32). Our task is made easier if we express the total wavefunction as a product of an electronic wavefunction and a nuclear one. For the present purpose it is sufficient to include only vibrational motion in the nuclear wavefunction. We now have

$$R_{e'V'e''V''} = \int \psi_e'^* \psi_V'^* (\hat{M}_e + \hat{M}_n) \psi_e'' \psi_V'' \, d\tau_e \, d\tau_V \tag{5.19}$$

where $\hat{M}_e$ and $\hat{M}_n$ are the dipole operators for the electrons and nuclei respectively. Expanding this integral, and noting that the nuclear dipole moment term is independent of electronic coordinates, yields the two integrals

$$R = \int \psi_e'^* \psi_V'^* \hat{M}_e \psi_e'' \psi_V'' \, d\tau_e \, d\tau_V + \int [\psi_e'^* \psi_e'' \, d\tau_e] \psi_V'^* \hat{M}_n \psi_V'' \, d\tau_V$$

The orthogonality of the electronic wavefunctions renders the second integral zero. In order to make a further simplification to the first part, we must invoke the Franck–Condon principle. This is really an application of the Born–Oppenheimer approximation to an electronic transition. The principle states that *an electronic transition takes place so rapidly compared to the vibrational motion of the nuclei that the internuclear distance can be regarded as fixed during the transition.* If this is the case the transition moment integral reduces to

$$R = \bar{R}_{e'e''} \int \psi_V'^* \psi_V'' \, d\tau_V \tag{5.20}$$

where $\bar{R}_{e'e''} = \int \psi_e'^* \hat{M}_e \psi_e'' \, d\tau_e$ is the *average* electronic transition moment. Since the transition intensity is proportional to the square of

Table 5.3 Spectroscopic constants for low-lying electronic states of several diatomic molecules

Molecule	state	T_e	$\overline{\omega_e}$	$\overline{\omega_e x_e}$	$\bar{B}_e$	$\bar{\alpha}_e$	r_e
CO	$X\,^1\Sigma^+$	0	2169.8	13.29	1.9313	0.0175	1.1283
	$A\,^1\Pi$	65075.7	1518.2	19.4	1.6115	0.0232	1.2353
	$B\,^1\Sigma^+$	86945.2	2112.7	15.2	1.9612	0.0261	1.1197
CN	$X\,^2\Sigma^+$	0	2068.6	13.09	1.8997	0.0174	1.1718
	$A\,^2\Pi$	9245.3	1812.5	12.60	1.7151	0.0171	1.2333
	$B\,^2\Sigma^+$	25752.0	2163.9	20.2	1.973	0.023	1.150
BeO	$X\,^1\Sigma^+$	0	1487.3	11.83	1.6510	0.0190	1.3309
	$A\,^1\Pi$	9405.6	1144.2	8.415	1.3661	0.0163	1.4631
	$B\,^1\Sigma^+$	21253.94	1370.8	7.746	1.5758	0.0154	1.3623
N_2	$X\,^1\Sigma_g^+$	0	2358.6	14.32	1.9982	0.0173	1.0977
	$A\,^3\Sigma_u^+$	50203.6	1460.6	13.87	1.4546	0.0180	1.2866
	$B\,^3\Pi_g$	59619.3	1733.4	14.12	1.6374	0.0179	1.2126
	$B'\,^3\Sigma_u^-$	66272.4	1516.88	12.18	1.473	0.0166	1.278

Note: The letters prefixed to the states are used to distinguish between states of the same type. The ground state is given the symbol X

the transition moment integral we find the important result that:

$$I_{e'V'e''V''} \alpha |\int \psi_V'^{*} \psi_V'' \, d\tau_V|^2 \qquad (5.21)$$

that is the intensity depends on the square of the overlap of the vibrational wavefunctions. In order for the integral in Equation (5.21), called a *Franck–Condon factor*, to have a significant value, it requires that the vibrational wavefunctions in both states must have large probabilities at the same values of internuclear distance.

If both electronic states had exactly the same shape and equilibrium bond length then Equation (5.21) would result in a 'rule' of $\Delta V=0$. The near ultra-violet absorption spectrum of the CN radical ($^2\Sigma^+ \leftarrow {}^2\Sigma^+$) consists of a 0–0 band and 1–0 band with a relative intensity of 11:1. On the above argument we would thus expect very similar electronic curves—which is confirmed by the similarity of the vibrational frequencies and bond lengths in each electronic state (see *Table 5.3*).

Figure 5.6 illustrates the application of the Franck–Condon principle for three typical cases.

No attempt will be made here to evaluate vibrational intensities via Equation (5.21), although the reader is directed to Problem 5.11 for a special case. We can, however, evaluate the positions of the vibrational lines easily from Equation (5.17) and this is done in Program 5.3. The vibrational line spacing information is most simply expressed as a *Deslandres table*, which places the lower vibrational quantum number across the top and the upper vibrational quantum

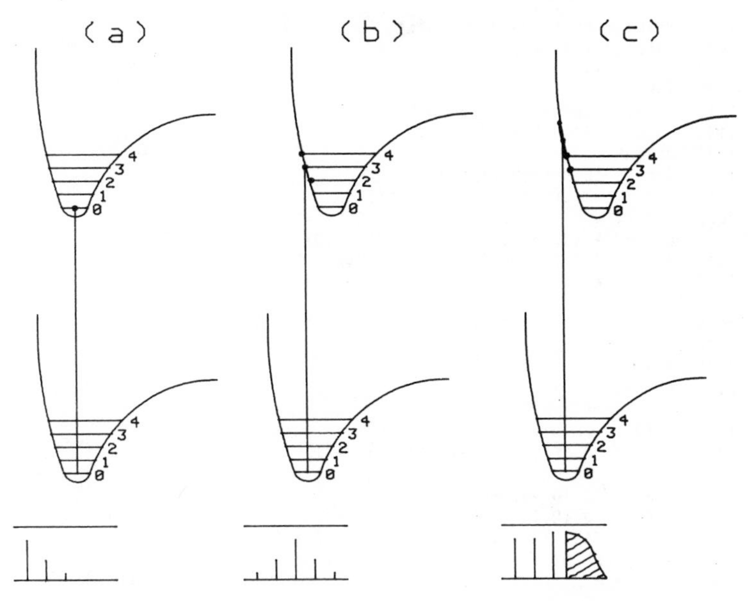

Figure 5.6 The spectra arising from the application of the Franck–Condon principle where the equilibrium bond length in the upper state is (a) similar to the ground state, (b) a little larger than the ground state, (c) much larger than the ground state

number down one side. Such tables are very useful when analysing unknown spectra, although many entries may be absent or weak because of low Franck–Condon factors. *Table 5.3* provides the required spectroscopic constants for some electronic states of a few diatomic molecules. More complete tables are available in Reference 2.

Program 5.3 DESTAB: Diatomic Deslandres table

```
90  DIM NU(10,10)
100  REM  DESLANDRES TABLE
110  GOSUB 1000: REM  GET DATA
120  GOSUB 2000: REM  CALCULATE TABLE
130  GOSUB 3000: REM  PRINT TABLE
140  END
1000  REM  INPUT
1010  PRINT : INPUT "MOLECULE = ";MN$
1020  PRINT : PRINT "ALL VALUES IN cm-1"
1030  PRINT : PRINT "LOWER STATE"
1040  INPUT "Te = ";TL
1050  INPUT "We = ";WL
1060  INPUT "WeXe = ";XL
```

```
1070  PRINT : PRINT "UPPER STATE"
1080  INPUT "Te = ";TU
1090  INPUT "We = ";WU
1100  INPUT "WeXe = ";XU
1110  PRINT : INPUT "MAX LOWER V = ";VL
1120  INPUT "MAX UPPER V = ";VU
1130  RETURN
2000  REM  CALC ENERGY DIFFERENCE
2010 TE = TU - TL: REM  ELEC DIFFERENCE
2020  FOR V1 = 0 TO VL: REM  LOWER LEVEL
2030 GL = WL * (V1 + .5) - XL * ((V1 + .5) ^ 2)
2040  FOR V2 = 0 TO VU: REM  UPPER LEVEL
2050 GU = WU * (V2 + .5) - XU * ((V2 + .5) ^ 2)
2060 GV = GU - GL
2070 NU(V2,V1) =  INT ((TE + GV) * 10 + .5) / 10
2080  NEXT
2090  NEXT
2100  RETURN
3000  REM  PRINT RESULTS
3010  PRINT : PRINT "   V''";
3020  FOR I = 0 TO VL: PRINT  TAB( (I + 1) * 9);I;: NEXT
3030  PRINT : PRINT "V'"
3040  FOR I = 0 TO VU: PRINT I;
3050  FOR J = 0 TO VL
3060  PRINT  TAB( (J + 1) * 9 - 4);NU(I,J);
3070  NEXT : PRINT : NEXT
3080  RETURN
```

```
MOLECULE = CO

ALL VALUES IN cm-1

LOWER STATE
Te = 0
We = 2169.8
WeXe = 13.29

UPPER STATE
Te = 65075.7
We = 1518.2
WeXe = 19.4

MAX LOWER V = 3
MAX UPPER V = 3

   V''  0         1         2         3
V'
0    64748.4  62605.2  60488.5  58398.5
1    66227.8  64084.6  61967.9  59877.9
2    67668.4  65525.2  63408.5  61318.5
3    69070.2  66927    64810.3  62720.3
```

5.8 Rotational fine structure

Unlike vibration the rotational transitions are still subject to selection rules arising from the conservation of angular momentum. In order to apply these rules we must first classify the rotational levels by their parity and symmetry as described in Section 5.3. The selection rules are:

$$\Delta J = 0, \pm 1 \qquad (J=0 \not\leftrightarrow J=0)$$

$$\Delta J \neq 0 \quad \text{for} \quad \Sigma \leftrightarrow \Sigma \tag{5.2}$$

$$+ \leftrightarrow - \qquad + \not\leftrightarrow + \qquad - \not\leftrightarrow -$$

$$\text{s} \leftrightarrow \text{s} \qquad \text{a} \leftrightarrow \text{a} \qquad \text{s} \not\leftrightarrow \text{a}$$

The positions of the rotational lines are given by

$$\Delta F = B'_{\text{V}} J'(J'+1) - B''_{\text{V}} J''(J''+1) \tag{5.23}$$

We can express the P, Q, R branches arising from Equation (5.23) by just two equations

$$\begin{aligned} \Delta F_{\text{PR}} &= (B'+B'')m + (B'-B'')m^2 \qquad m = \pm 1, \pm 2 \ldots \\ \Delta F_{\text{Q}} &= (B'-B'')m + (B'-B'')m^2 \qquad m = 1, 2 \ldots \end{aligned} \tag{5.24}$$

where $m = J''+1$ for the R branch and $m = -J''$ for the P branch. In all three cases m cannot take the value of zero. This leads to the so-called *zero gap* at the pure vibrational frequency. The expressions in Equation (5.24) are the equations of parabolae and are referred to as *Fortrat Parabolae*. *Figure 5.7* shows the rotational spectrum accompanying the 0–0 vibrational transition in the $B^2\Sigma^+ \leftarrow X^2\Sigma^+$ absorption spectrum of CN. The Fortrat parabola is also drawn through the data points.

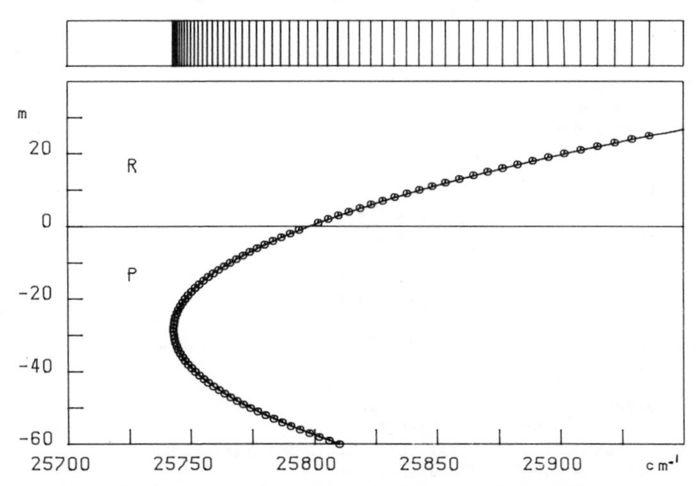

Figure 5.7 The rotational structure accompanying the 0–0 band of the $B\,^2\Sigma^+ \leftarrow X\,^2\Sigma^+$ electronic transition of CN. The upper panel shows the line spectrum

The spectrum has no Q branch since this is not allowed in $\Sigma \leftarrow \Sigma$ transitions. The position of the missing line at $m=0$ is called the *band origin.* In fact, since lines appear to the left and right of this value, it is not a simple matter to determine this quantity. The experimental spectrum appears to start at a value corresponding to the vertex of the Fortrat parabola—in this case at a value of $m=-28$ in the P branch. The vertex of the P, R parabola is called the *band head* and in general it can lie in either the P or the R branch. We can calculate the position of the vertex by differentiating Equation (5.24) and equating the result to zero.

$$\frac{\mathrm{d}}{\mathrm{d}m}(\Delta F_{\mathrm{PR}})=(B'+B'')+2(B'-B'')m=0$$

$$m=\frac{(B'+B'')}{2(B'-B'')} \tag{5.25}$$

The Fortrat parabola is a very compact way of describing the apparently complex series of lines that make up the rotational fine structure. Program 5.4 uses the data of *Table 5.3* to calculate the P, Q, R rotational spectrum accompanying a particular vibrational line of an electronic transition. No attempt is made here to calculate the line intensities, nor are the selection rules incorporated, so the Q branch is included regardless of whether it can occur.

Program 5.4 ROTFIN: Rotational fine structure

```
90  DIM DA(2,6)
100  REM  ROTATIONAL FINE STRUCTURE
110  GOSUB 1000: REM  GET DATA
120  GOSUB 2000: REM  CALCULATE CONSTANTS
130  GOSUB 3000: REM  CALCULATE P,Q,R BRANCHES
140  END
1000  REM  INPUT
1010  PRINT : INPUT "MOLECULE = ";MN$
1020  PRINT : PRINT "ALL VALUES IN cm-1": PRINT
1030  FOR I = 1 TO 2
1040  IF I = 1 THEN  PRINT "LOWER STATE"
1050  IF I = 2 THEN  PRINT "UPPER STATE"
1060  PRINT "Te,We,WeXe,Be,Ae,V "
1070  INPUT DA(I,1),DA(I,2),DA(I,3),DA(I,4),DA(I,5),DA(I,6)
1080  NEXT
1090  PRINT : INPUT "MAX J LEVEL = ";JM
1100  RETURN
2000  REM  CALC ENERGY DIFFERENCE
2010 TE = DA(2,1) - DA(1,1): REM  ELECTRONIC DIFFERENCE
2020  REM  NOW VIBRATIONAL DIFFERENCE
2030 GL = DA(1,2) * (DA(1,6) + .5) - DA(1,3) * ((DA(1,6) + .5) ^ 2)
2040 GU = DA(2,2) * (DA(2,6) + .5) - DA(2,3) * ((DA(2,6) + .5) ^ 2)
2050 GV = GU - GL
2060 NU = TE + GV: REM  EL AND VIB FREQUENCY
```

```
2070  REM  NOW ROT CONSTANTS
2080 BL = DA(1,4) - DA(1,5) * (DA(1,6) + .5)
2090 BU = DA(2,4) - DA(2,5) * (DA(2,6) + .5)
2100 BP = BU + BL:BM = BU - BL
2110  RETURN
3000  REM  P,Q,R BRANCHES
3010  PRINT "J     P"; TAB( 14)"J     Q"; TAB( 28);"J     R"
3020  FOR M = 1 TO JM
3030 FR = BP * M + BM * M * M + NU
3040 FP =  - BP * M + BM * M * M + NU
3050 FQ = BM * M + BM * M * M + NU
3060 FR =  INT (FR * 10 + .5) / 10:FP =  INT (FP * 10 + .5) / 10:FQ =  INT
    (FQ * 10 + .5) / 10
3070  PRINT M;"  ";FP; TAB( 14);M;"  ";FQ; TAB( 28);M - 1;"  ";FR
3080  NEXT
3090 MV =  - BP / (2 * BM): REM  VERTEX
3100 NV = NU + BP * MV + BM * MV * MV
3110  PRINT : PRINT "BAND ORIGIN = "; INT (NU * 10 + 0.5) / 10;" cm-1"
3120  PRINT "BAND HEAD   = "; INT (NV * 10 + .5) / 10;" cm-1";
3130  IF MV < 0 THEN  PRINT "  (P BRANCH)"
3140  IF MV > 0 THEN  PRINT "  (R BRANCH)"
3150  RETURN
```

```
MOLECULE = CN

ALL VALUES IN cm-1

LOWER STATE
Te,We,WeXe,Be,Ae,V
?0,2068.6,13.09,1.8997,0.0174,0
UPPER STATE
Te,We,WeXe,Be,Ae,V
?25752,2163.9,20.2,1.973,0.023,0

MAX J LEVEL = 35
J     P        J    Q         J     R
1  25794.1    1  25798       0  25801.8
2  25790.4    2  25798.3     1  25805.9
3  25786.9    3  25798.7     2  25810.1
4  25783.6    4  25799.3     3  25814.4
5  25780.4    5  25800       4  25818.9
6  25777.3    6  25800.8     5  25823.5
7  25774.4    7  25801.8     6  25828.3
8  25771.6    8  25802.9     7  25833.2
9  25768.9    9  25804.2     8  25838.3
10  25766.4   10  25805.6    9  25843.4
11  25764     11  25807.2    10  25848.8
12  25761.8   12  25808.9    11  25854.3
13  25759.7   13  25810.7    12  25859.9
14  25757.8   14  25812.7    13  25865.6
15  25755.9   15  25814.8    14  25871.5
16  25754.3   16  25817      15  25877.6
17  25752.8   17  25819.4    16  25883.7
18  25751.4   18  25822      17  25890.1
19  25750.1   19  25824.7    18  25896.5
20  25749     20  25827.5    19  25903.1
21  25748.1   21  25830.4    20  25909.9
22  25747.2   22  25833.5    21  25916.7
23  25746.6   23  25836.8    22  25923.8
```

```
24  25746     24  25840.2   23  25930.9
25  25745.6   25  25843.7   24  25938.2
26  25745.4   26  25847.4   25  25945.7
27  25745.2   27  25851.2   26  25953.3
28  25745.3   28  25855.1   27  25961
29  25745.4   29  25859.2   28  25968.9
30  25745.7   30  25863.4   29  25976.9
31  25746.2   31  25867.8   30  25985.1
32  25746.8   32  25872.3   31  25993.3
33  25747.5   33  25877     32  26001.8
34  25748.4   34  25881.8   33  26010.4
35  25749.4   35  25886.7   34  26019.1

BAND ORIGIN = 25797.9 cm-1
BAND HEAD   = 25745.2 cm-1   (P BRANCH)
```

The rotational spectra of homonuclear diatomic molecules are of course influenced by nuclear spin just as we saw in Section 5.3 and these effects must be included in any description of line intensities. In fact the rotational fine structure exhibits even more complex behaviour when the interactions between molecular rotation, electronic orbital angular momentum and electron spin are included. In some cases this leads to early rotational lines being entirely absent from the spectrum ($J=0$ for Π states, $J=0$, 1 for Δ states, etc.). Higher resolution reveals that the rotational lines are themselves split by these interactions and each rotational line is made up of several closely spaced lines. The reader should consult the books listed at the end of this chapter for further details.

5.9 Electronic spectra of polyatomic molecules

It will be appreciated from the foregoing section that the treatment of the high resolution electronic spectroscopy of polyatomic molecules is beyond the scope of an introductory book. Although the study of low resolution electronic spectra is more feasible it requires a certain fluency with the concepts of molecular symmetry and the techniques of group theory, in order to make reasonable progress. There is, however, one class of molecules for which a simple model provides a surprisingly good description of the observed electronic spectra. It is this class, the linear polyenes, to which we now turn our attention.

5.9.1 *Electronic spectra of conjugated polyenes*

The π electrons in a fully conjugated linear polyene (all carbon atoms are sp^2 hybrids) are best thought of as being delocalized over the whole molecule. In this case we can (roughly) replace the actual potential experienced by the π electrons with a uniform potential along the carbon skeleton. If we also decide to ignore the fact that the carbon atoms zig-zag along the chain, then our description reduces to that of electrons confined to a one-dimensional box. As a final drastic simplification, we will further assume that the electrons do not interact with each other apart from the restriction imposed by the Pauli principle. These approximations constitute the *free electron model*. To apply it we only require the solutions to the Schrödinger equation for a particle in a one-dimensional box. For a box of length L we find wavefunctions given by:

$$\psi = \sqrt{\frac{2}{L}} \sin \frac{n\pi x}{L} \qquad (5.26)$$

with energy levels

$$E_n = \frac{n^2 h^2}{8mL^2} \quad \text{J} \qquad (5.27)$$

The π electrons are allocated to the energy levels of Equation (5.27), bearing in mind the Pauli principle which requires that only two electrons be placed in each level and that these must have opposite spins. *Figure 5.8* shows this for the molecule 1,3,5 hexatriene; also indicated is the electronic transition corresponding to the first UV absorption band of this molecule.

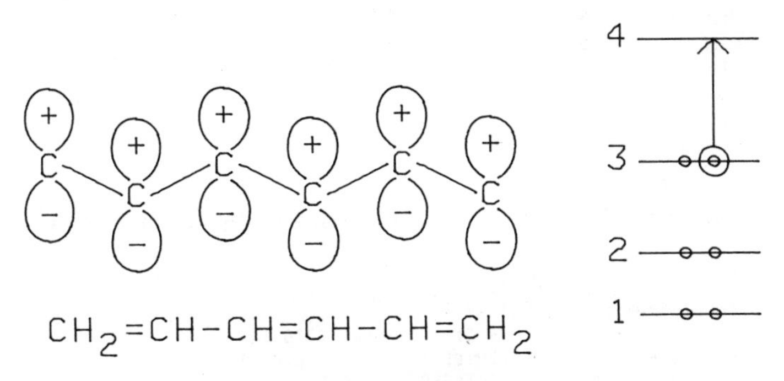

Figure 5.8 The free electron model for 1,3,5 Hexatriene

The selection rule for absorption is easily seen to be $\Delta n = +1, +3 \ldots$ from symmetry considerations (see Problem 5.14). If n is the highest occupied level, then the energy spacing for the lowest electronic transition is simply given by

$$\Delta E = E_{n+1} - E_n = \frac{h^2}{8mL^2}(2n+1) \quad \text{J} \qquad (5.28)$$

This corresponds to the smallest energy difference and hence the longest wavelength at which the transition will be observed. In fact, the absorption bands are somewhat wide and the experimental values quoted are usually for λ_{max}—the wavelength at maximum intensity.

The length of the box will not be equal to the length of the carbon chain since the electron density will extend a little way past the ends of the molecule. An empirical relationship is used here which works well at reproducing λ_{max} values[3]

$$L = 1.4(k+2) \qquad \text{angstroms} \tag{5.29}$$

where k is the number of formal double bonds and 1.4 Å is the average C–C bond length. The use of an empirical approach for determining L enables us to compensate to some extent for the weaknesses of the free electron model.

Program 5.5 performs the calculations described above for fully conjugated linear polyenes. *Table 5.4* shows the results for the first four molecules in the series.

Table 5.4 Results of the free electron model applied to linear polyenes

Molecule	N_c	k	L	λ_{max} (calc)	λ_{max} (exp)
Ethylene	2	1	4.2	193.9	162.5
Butadiene	4	2	5.6	206.8	210.0
Hexatriene	6	3	7.0	230.8	247.0
Octatetraene	8	4	8.4	258.5	286.0

Note: λ in nm and L in Å

Program 5.5 FEMLIN: Free electron model for linear polyenes

```
100  REM  FREE ELECTRON MODEL
110  INPUT "NUMBER OF C ATOMS = ";N
120  IF (N -  INT (N / 2) * 2) = 0 THEN  GOTO 140
130  PRINT "NOT ALL sp2 HYBRIDS": PRINT : GOTO 110
140 K = N / 2: REM  TWO ELECTRONS PER LEVEL
150  REM  NOW LENGTH OF BOX
160 L = 1.4 * (K + 2): REM  ANGSTROMS
170  PRINT : PRINT "BOX LENGTH = "; INT ((L * 10 + .5)) / 10;" ANGSTROMS"

180  REM  NOW ENERGY LEVELS
190  PRINT : PRINT "ENERGY LEVEL INFORMATION": PRINT
200  FOR I = 1 TO K + 1
210 E = (37.5897 * I * I) / (L * L): REM  eV
220 E =  INT (E * 1000 + .5) / 1000: REM  ROUND FIGURES
230  PRINT I;"   ";
240  IF I <  = K THEN  PRINT E; TAB( 12);"eV   (OCCUPIED)"
250  IF I = K + 1 THEN  PRINT E; TAB( 12);"eV  (UNOCCUPIED)"
```

```
260  NEXT
270 LA = (32.972968 * L * L) / (2 * K + 1): REM  LAMBDA IN nm
280 LA =  INT (LA * 10 + .5) / 10
290  PRINT : PRINT "WAVELENGTH = ";LA;"  nm"
```

```
NUMBER OF C ATOMS = 10

BOX LENGTH = 9.8 ANGSTROMS

ENERGY LEVEL INFORMATION

1  .391     eV    (OCCUPIED)
2  1.566    eV    (OCCUPIED)
3  3.523    eV    (OCCUPIED)
4  6.262    eV    (OCCUPIED)
5  9.785    eV    (OCCUPIED)
6  14.09    eV  (UNOCCUPIED)

WAVELENGTH = 287.9  nm
```

Program note

(1) The energy levels are given to show relative spacings only—they are not accurate on an *absolute* basis.

The agreement with experiment is remarkably good considering the drastic simplifications that are involved. This agreement is maintained even with very long chains. For instance the molecule β *carotene*, a yellow pigment found in carrots and green leaves, contains 22 conjugated carbon atoms and has a predicted λ_{max} of 475 nm. This compares favourably with its experimental value of 451 nm at the violet end of the spectrum. The model has also been applied to heteroatom systems such as the cyanine dye cations and to calculating other electronic properties such as charge densities. The interested reader should refer to the excellent discussion in Reference 3 for more detail about the treatment of π electron systems.

5.10 References

1. Herzberg, G., *Spectra of Diatomic Molecules*, Van Nostrand (1950).
2. Huber, K.P. and Herzberg, G., *Constants of Diatomic Molecules*, Van Nostrand (1979).
3. Pilar, F.L., *Elementary Quantum Chemistry*, McGraw-Hill (1968).

5.11 Further reading

1. Banwell, C.N., *Fundamentals of Molecular Spectroscopy*, McGraw-Hill (1972).
2. Dixon, R.N., *Spectroscopy and Structure*, Methuen & Co. (1965).
3. Herzberg, G., *The Spectra and Structure of Simple Free Radicals*, Cornell University Press (1971).
4. Herzberg, G., *Infra-red and Raman Spectra*, Van Nostrand (1945).
5. Steinfeld, J.I., *Molecules and Radiation*, The MIT Press (1974).

PROBLEMS

(5.1) Use the data of *Table 4.3* to calculate the intensities and shifts of the vibrational Stokes and Anti-Stokes lines in H_2, HCl, N_2 and I_2.
(5.2) In the pure rotational Raman spectrum of a diatomic molecule, the distance (from the exciting line) of the first Stokes and Anti-Stokes line is different from the subsequent line spacing within each branch. Calculate the ratio of these spacings when, (i) all rotational levels are populated, (ii) only the odd levels are populated and (iii) only the even levels are populated.
(5.3) The Raman spectrum of $^{16}O_2$ reveals a ratio of 5:2 for the line spacings described in the previous problem. Given that the electronic ground state of O_2 is $^3\Sigma_g^-$, what is the nuclear spin of ^{16}O?
(5.4) Modify Program 5.1 to calculate the vibration–rotation Raman spectrum. (Hint: you will need the Boltzmann population for the initial V, J level.)
(5.5) Modify Program 5.2 to require only the chemical symbol and the isotopic mass at the input stage. (Hint: you will need the data of *Table 5.1* stored internally.)
(5.6) Modify Program 5.1 to include the effect of nuclear spin statistics by combining Programs 5.1 and 5.2.
(5.7) The linear molecule N_2O exhibits strong infra-red and Raman lines at 588.8 cm^{-1} (IR), 1285 cm^{-1} (IR and R) and 2223.5 cm^{-1} (IR and R). Is the molecular structure NON or NNO?
(5.8) Draw the polarizability ellipsoids for the vibrational modes of H_2O (see *Figure 4.8*). Explain why all three modes are Raman active (although only v_1 is very strong). (Hint: remember for v_2 that a change in *direction* of the polarizability ellipsoid will lead to activity.)
(5.9) The polarizability and momental axes of symmetric top molecules coincide for symmetry reasons. This leads to selection rules for *rotational* Raman scattering of the form

$$\Delta J = 0, \pm 1, \pm 2 \qquad \Delta K = 0$$

$$\Delta J \neq \pm 1 \qquad \text{if } K = 0$$

Use Equation (3.33) or (3.34) to evaluate the rotational Raman line spacings for symmetric top molecules. (Remember only positive values of ΔJ are possible.)
(5.10) What electronic transitions are possible starting from the following states $^1\Delta_g$, $^2\Pi_u$, $^1\Sigma^-$?
(5.11) Modify the program of Problem 4.4 to calculate the Franck–Condon factors, for the first three vibrational levels, between two (simple harmonic) electronic states of identical shape making use of Equation (5.21). Use the program to calculate the vibrational coarse

structure for $r'_e = r''_e$, $r'_e \sim 1.5r''_e$ and $r'_e = 2r''_e$. (Hint: you will need the wavefunctions of Section 4.1.5.)

(5.12) Show that Equation (5.24) follows from Equations (5.23) and (5.22) and that $m=0$ is not allowed by the selection rules.

(5.13) If the rotational fine structure of an electronic transition has its band head in the P branch, is the equilibrium bond length in the upper state longer or shorter than in the lower state?

(5.14) Show that the selection rule for the particle in the box is $\Delta n = \pm 1, \pm 3, \ldots$ by considering the odd and even properties of the wavefunctions and the dipole moment operator. (Hint: $\psi_m \hat{M}_e \psi_n$ must be even overall.)

(5.15) The free electron model can be extended to cyclic compounds like benzene by treating the π electrons as particles on a ring. The energy levels for this problem are given by

$$E_n = \frac{n^2 h^2}{8\pi^2 I} \qquad n = 0, \pm 1, \pm 2 \ldots$$

where I is the moment of inertia of the electron relative to the ring centre. The six π electrons in benzene occupy E_0, E_{-1} and E_1, hence the electronic transition is from $n=1$ to $n=2$. Calculate the wavelength of this transition and compare it to the experimental value of 205 nm. ($m_e = 9.11 \times 10^{-31}$ kg, radius of benzene $= 1.39$ Å.)

Appendix 1

A.1 Fundamental constants

The fundamental constants given here are all in S.I. units and their values should be used wherever their symbols appear in formulae. This applies even in the case where a formula produces a non-S.I. unit as a final result, such as cm^{-1}.

$c = 2.9979 \times 10^{8}$ m s^{-1}	Velocity of light
$k = 1.3805 \times 10^{-23}$ J K^{-1}	Boltzmann constant
$h = 6.6256 \times 10^{-34}$ J s	Planck constant
$e = 1.6021 \times 10^{-19}$ C	Charge on the electron
$L = 6.0225 \times 10^{23}$ mol^{-1}	Avogadro number
$\varepsilon_0 = 8.8542 \times 10^{-12}$ F m^{-1}	Permittivity of free space

A.2 Energy units

Table A.1 gives conversion factors between the commonly met energy units.

Examples of using the table would be: 1 cm^{-1} = 1.239×10^{-4} eV, or that to convert from Hz to kJ mol^{-1} we must multiply by 3.991×10^{-13}.

Table A.1 Conversion factors for commonly used energy units

	J	$kJ\ mol^{-1}$	Hz	cm^{-1}	eV
J	1	6.023×10^{20}	1.509×10^{33}	5.035×10^{22}	6.242×10^{18}
$kJ\ mol^{-1}$	1.66×10^{-21}	1	2.506×10^{12}	83.6	1.036×10^{-2}
Hz	6.626×10^{-34}	3.991×10^{-13}	1	3.336×10^{-11}	4.137×10^{-15}
cm^{-1}	1.986×10^{-23}	1.196×10^{-2}	2.998×10^{10}	1	1.239×10^{-4}
eV	1.602×10^{-19}	96.49	2.417×10^{-14}	8.066×10^{3}	1

Note: $1\ cm^{-1} = 100\ m^{-1}$

Index